T0391416

Hydrogen Membranes

This book provides an overview of the most recent advances and innovations in the subject of hydrogen separation, its applications, and the role of membranes in this process. It starts with the fundamental principles underlying hydrogen separation and the various types of membranes that are used for hydrogen separation. Furthermore, it explains different fabrication methods, including characterization techniques used to evaluate performance and properties. Finally, it covers diverse applications of hydrogen separation membranes backed by case studies and examples.

Features:

- Covers several forms of hydrogen membranes, such as polymeric, metallic, and ceramic membranes and provides a thorough understanding of their manufacturing, construction, and applications.
- Explores recent advancements in materials used for the fabrication of hydrogen membranes.
- Focuses on theoretical concepts and practical applications of hydrogen separation membranes.
- Explains the role of hydrogen membranes in the transition toward a sustainable energy future.
- Discusses real-world case studies, industrial applications, and potential future uses.
- Emphasizes real-world applications of hydrogen membranes in fields such as fuel cells, hydrogen generation, and carbon capture.

This book is aimed at academics, energy professionals, engineers, researchers, and scientists working on sustainable energy solutions. Readers will obtain a detailed grasp of hydrogen membrane technology, from basic principles to sophisticated applications, giving them the capacity to innovate further.

Hydrogen Membranes

Production, Fabrication, and Applications

Vamsi Krishna Kudapa and Surajit Mondal

CRC Press is an imprint of the
Taylor & Francis Group, an **informa** business

Designed cover image: Shutterstock

First edition published 2025
by CRC Press
2385 NW Executive Center Drive, Suite 320, Boca Raton FL 33431

and by CRC Press
4 Park Square, Milton Park, Abingdon, Oxon, OX14 4RN

CRC Press is an imprint of Taylor & Francis Group, LLC

ISBN: 9781032816975 (hbk)
ISBN: 9781032967950 (pbk)
ISBN: 9781003590682 (ebk)

DOI: 10.1201/9781003590682

Typeset in Times
by Newgen Publishing UK

Contents

Preface

The combustion of fossil fuels provides over 80% of the world's current energy demand. In contrast to the emissions of greenhouse gases caused by the combustion of fossil fuels, hydrogen combustion produces only water as a waste product. Hydrogen is a more environmentally friendly alternative fuel. The production of hydrogen energy has the potential to assist in addressing energy security issues such as climate change and air pollution. There is an increasing global interest in hydrogen, particularly green hydrogen, which is produced by electrolyzing water using power derived from renewable resources. Because of falling hydrogen prices and the growing urgency of decarbonization, global demand for hydrogen, headed by the transportation and industrial sectors, could increase by about 400% by 2050. Furthermore, using environmentally friendly hydrogen will result in a reduction of 3.6 gigatons of total CO_2 emissions between 2020 and 2050. Hydrogen has the highest energy density of any known fuel, and it is widely available in enormous quantities all over the planet. It is possible that by 2050, India's need for hydrogen will have increased by a factor of four, accounting for more than 10% of global consumption. Steel and heavy-duty transportation are expected to account for more than 52% of overall demand growth between now and 2050. The overall market value for environmentally friendly hydrogen in India might reach $8 billion by 2030 and $340 billion by 2050. Because India's capacity to create power from renewable sources is growing all the time, the country now can produce hydrogen from ecologically beneficial sources such as solar and wind when demand is low. Physical absorption and polymer membranes can be employed to extract hydrogen from crude hydrogen polluted with hydrocarbons. This can be done to clean the crude hydrogen. Furthermore, both hot alkaline absorption and conventional absorption are viable methods for removing CO_2 and H_2O from space. The purity of hydrogen is an important aspect in determining whether it can be used in the energy production process. Unlike other types of separation technologies, membrane processes can be used in both mobile and small-scale applications. The membrane may function properly under a wide range of pressure and temperature extremes. The fundamental objective and goal of the separation membrane is to be used in membrane reactors for synchronous hydrogen purification and production. Other competing methods, such as pressure swing adsorption and cryogenic distillation, do not compare favorably to the membrane separation approach. This is due to the ability of technology to separate compounds at considerably lower temperatures. This is because membrane separation takes fewer resources than other competing technologies, particularly ones that have been around for a longer time.

This book provides an overview of the importance of hydrogen separation, its applications, and the role of membranes in this process. The following topics will be covered in detail:

- The fundamental principles underlying hydrogen separation, including concepts like permeation, selectivity, diffusion, and solubility.

- The various types of membranes used for hydrogen separation, such as polymeric membranes, ceramic membranes, metallic membranes, and composite membranes. This book also explores their structures, properties, and performance characteristics.
- The different fabrication methods employed to produce hydrogen separation membranes, including solution casting, phase inversion, electrospinning, chemical vapor deposition (CVD), and physical vapor deposition (PVD).
- Various factors that influence the performance of hydrogen separation membranes, including temperature, pressure, feed composition, membrane thickness, and surface morphology.
- The diverse applications of hydrogen separation membranes, such as hydrogen production from natural gas, hydrogen purification, hydrogen recovery from industrial processes, and hydrogen fuel cells.
- The current challenges and limitations associated with hydrogen separation membranes, such as membrane fouling, stability, and cost.
- Real-world case studies and examples of hydrogen separation membrane applications in industries like petrochemicals.

About the Authors

Vamsi Krishna Kudapa has been Associate Professor in the Department of Chemical and Petroleum Engineering, UPES, Dehradun, since 2013. Before UPES, he worked as Assistant Professor and Head of the Department of Petroleum Engineering, Sri Aditya Engineering College, Andhra Pradesh. He pursued his Ph.D. in Oil & Gas—Modeling and Simulation from UPES, Dehradun, in 2018. He has 30 publications, two patents, and five book chapters to his credit. He is currently working on hydrogen energy and solar energy applications in agriculture, the application of nanoparticles in drilling fluids and cementation, enhanced oil recovery, and modeling and simulation of unconventional gas reservoirs.

Surajit Mondal completed his master's in Energy Systems and Ph.D. in the Renewable Energy domain in 2020. He teaches undergraduate, post-graduate, and Ph.D. scholars at the University of Petroleum and Energy Studies. He has published over 40 international research and review articles as an author/co-author. He has published 65 patents and granted 18 patents against his name. He has completed 2 DST (Govt. of India) funded projects of INR 1 crore approx. in the field of energy systems/sustainability.

1 Introduction to Hydrogen and Hydrogen Separation

1.1 INTRODUCTION

Hydrogen energy has emerged as a promising solution to tackle numerous urgent challenges facing the world today. The widespread use of fossil fuels for energy is swiftly leading to significant environmental issues globally. In our efforts to combat climate change, decrease dependency on fossil fuels, and transition toward sustainable energy sources, hydrogen presents a compelling alternative. Its adaptability, abundance, and potential for emitting zero pollutants position it as a vital component in the realm of clean energy. A primary advantage of hydrogen energy lies in its capacity to decarbonize various sectors, particularly transportation and industry. Hydrogen fuel cells can propel vehicles, providing an alternative to traditional internal combustion engines and diminishing greenhouse gas emissions. Furthermore, hydrogen finds utility in industrial processes like steel and chemical production, where reducing carbon output is challenging but essential for curbing overall emissions. Addressing the inconsistency of renewable energy sources like solar and wind, hydrogen emerges as a promising solution for storing energy. It can capture excess power generated during peak production times and release it back into the grid, resulting in strengthening energy security. Moreover, hydrogen presents opportunities for energy independence and diversification, as it can be derived from various sources, including renewable electricity, natural gas with carbon capture and storage (CCS), biomass, and even water through electrolysis. This multiplicity of production methods reduces reliance on imported fossil fuels and strengthens energy resilience. The idea of a hydrogen-based economy, where hydrogen fuels our lives, first surfaced in 1970. Presented by John Bockris at General Motors, this "hydrogen economy" envisioned using hydrogen for everything—from generation (through fossil fuels, renewables, or even electricity) to distribution and everyday use. The concept gained momentum after the global energy crisis of 1974, highlighting the need for alternative energy sources. Despite not being readily available in its pure form, hydrogen's exceptional properties render it an exceedingly promising energy carrier or fuel. Various methods are currently employed for large-scale hydrogen production. With its widespread availability, hydrogen can be derived from diverse materials and compounds, utilizing a broad spectrum of methods, including environmentally friendly approaches. Crucially, hydrogen production is feasible worldwide.

DOI: 10.1201/9781003590682-1

1.2 PROPERTIES AND CHARACTERISTICS OF HYDROGEN AS TOMORROW'S FUEL

The simplest and lightest element in the periodic table, hydrogen has atomic number one, which qualifies it for usage in balloons. Since it makes up 75% of the mass of the cosmos, it is also the most plentiful element. Owing to its reactivity, it usually combines with other elements to form compounds or diatomic molecules (H_2). It exists as an odorless, colorless molecule with a density one-fourteenth that of air at typical ambient temperature and pressure (McCay and Shafiee, 2020). Due to its unique properties, hydrogen stands out as a versatile and promising fuel. Unlike traditional fuels derived from petroleum or natural gas, hydrogen's tiny, lightweight molecules offer significant advantages. Hydrogen's potential as a clean and versatile fuel stems from its unique properties. Unlike bulky molecules found in conventional fossil fuels like gasoline or natural gas, hydrogen boasts a remarkably small and lightweight structure (Mazloomi and Gomes, 2012). With its lightweight nature and high energy-to-weight ratio, hydrogen serves as an optimal energy carrier and boasts numerous properties that position it as a promising fuel for diverse applications. Three key properties underline hydrogen's potential as a future energy solution (Figure 1.1).

To begin with, hydrogen is abundant and can be obtained from sources such as water or hydrocarbons. When utilized in fuel cells, hydrogen generates electricity cleanly, emitting only water vapor as a byproduct. Its high energy density by weight makes it well-suited for transportation applications, where weight plays a crucial role. Hydrogen's versatility allows for its utilization across various sectors, including transportation, electricity generation, industrial processes, and heating. Electrolysis, utilizing renewable energy sources, can produce hydrogen, contributing to efforts aimed at reducing carbon emissions. Different storage methods such as compressed gas and liquid form provide flexibility for diverse applications. National Aeronautics and Space Administration (NASA) has been using liquid hydrogen to launch rockets, including the space shuttle, into orbit since the 1970s. The shuttle's

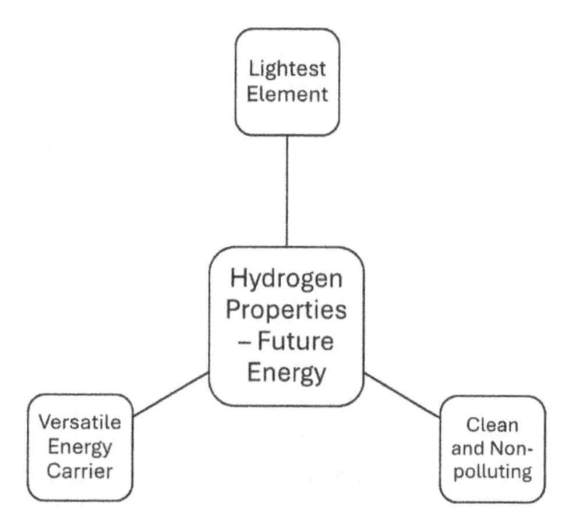

FIGURE 1.1 The three key aspects of utilizing hydrogen as a fuel source in the future.

TABLE 1.1
Hydrogen's Characteristics: A Fuel for Tomorrow

Property	Value
Name and symbol	Hydrogen and H
Periodic table number	1
Category	Non – mental
Atomic weight	1.00784
Color, odor	Colorless, odorless
Phase	Gas
Electrons, protonts, and neutrons	1,1,0
Toxicity	None
Liquid – to – gas expansion ration	1:848 at atmospheric conditions
Melting and boiling point	14.1 K and 20.28 K
Isotopes	3
Ionic radius	0.208 nm
Yander Waals radius	0.12 nm
Relative desity (concerning air)	0.0695 gm/1
Critical temperature	33.15 K
Critical pressure	13.0 atmospheres

TABLE 1.2
The Energy Densities of Different Fuels by Weight and Volume

Fuel	Gravity Energy Density (MJ/kg)	Volum e Energy Density (MJ/1)
Hydrogen	143	10.1
Methane (atmospheric pressure)	55.6	0.0378
Natural gas	53.6	0.0364
Natural gas (compressed)	53.6	9
Natual gas (liquid)	53.6	22.2
LPG propane	49.6	25.3
LPG butane	49.1	27.7
Petrol	46.4	34.2
Biodiesel	42.2	33
Diesel	45.4	34.6

electrical systems are powered by hydrogen fuel cells, which produce pure water as a clean byproduct that the crew can drink (Mazloomi and Gomes, 2012). Nevertheless, it is crucial to prioritize safety and address concerns related to its flammability and appropriate handling. Table 1.1 provides a summary of certain properties associated with hydrogen molecules.

The hydrogen nucleus comprises one proton and one electron. Bombarding hydrogen-containing compounds with neutrons can result in the formation of hydrogen isotopes (Winter, 1999), namely deuterium (H_2) and tritium (H_3). These isotopes possess radioactive properties, and nuclear devices have been constructed and tested using these materials (Momirlan and Veziroglu, 2005). Hydrogen exhibits an energy content per unit mass of 143 MJ/kg, which is up to three times greater than that of liquid hydrocarbon-based fuels (Serpone et al., 1992). Table 1.2 illustrates the volumetric and gravimetric energy density of hydrogen compared to other commonly used fuels (Momirlan and Veziroglu, 2005).

1.3 IMPORTANCE OF HYDROGEN IN CLEAN ENERGY SYSTEMS—COLOR CODES

Hydrogen holds immense promise as a fundamental component of a forthcoming energy system characterized by cleanliness, security, and cost efficiency. Its versatility allows for enhanced integration of renewables like wind and solid photovoltaic (PV) into the energy mix. A pivotal aspect of integrating hydrogen into energy strategies lies in its capacity to facilitate the decarbonization of the energy sector by offering a clean and dependable source for energy storage and transportation. However, the production of hydrogen from low-carbon sources remains prohibitively costly, with most of the hydrogen currently sourced from natural gas and coal. Present regulations present obstacles to the advancement of a clean hydrogen industry (Arcos and Santos, 2023).

The production of hydrogen relies heavily on the technology and source employed, with nearly 13 distinct color codes used to denote hydrogen production from various sources. Green, blue, grey, brown, black, turquoise, purple, pink, red, and white are among the color codes. In the energy sector, identifiers, or color codes such as green, blue, brown, yellow, turquoise, and pink, are frequently used to distinguish between different hydrogen sources. Hydrogen can have varied colors depending on how it is produced. However, interpretations might change over time and between nations because there isn't a universal naming convention. The generation of brown, grey, and black hydrogen significantly increases greenhouse gas (GHG) emissions. Most of these hydrogen types come from fossil fuels like coal and natural gas. During the process of producing grey hydrogen, which is mostly produced by steam methane reforming (SMR), carbon dioxide is released. The use of bituminous (black) or lignite (brown) coal results in the production of either black or brown hydrogen during the coal-using process. Despite producing hydrogen, the process of coal gasification is very harmful as it releases carbon dioxide and carbon monoxide into the atmosphere (Ishida et al., 2017).

While the proportion of grey and black hydrogen is declining, SMR, which releases carbon dioxide into the environment, is still a major source of brown hydrogen production in the US. While hydrogen fuel cells offer a solution to reduce air pollution, relying on fossil fuels for hydrogen production isn't sustainable. A green hydrogen future depends on developing technologies that can produce hydrogen without releasing GHGs into the atmosphere. Even though producing blue hydrogen is more environmentally friendly than using conventional techniques, it nevertheless produces CO_2 and falls short of global GHG emission reduction targets. In contrast, green hydrogen, produced from renewable energy sources like solar panels and wind turbines through water electrolysis, aims for zero emissions, significantly reducing carbon footprint. However, the cost of producing green hydrogen is now higher, in part because of the price of the materials used to make electrolyzers (Grid, 2023).

While the production of hydrogen from renewable sources has not advanced very much, current developments show promising results. Green hydrogen is an alternative that produces no carbon emissions by the electrolysis of water using renewable energy. On the other hand, carbon neutrality is achieved by blue hydrogen, which is made from fossil fuels but contains sequestered carbon dioxide that is stored underground. To make use of trapped carbon, some businesses investigate carbon capture, storage, and use (CCSU). Methane pyrolysis yields turquoise hydrogen, which transforms carbon into a solid state while using a lot less energy than water electrolysis. Nuclear-generated hydrogen includes pink hydrogen produced by nuclear-powered water electrolysis and purple hydrogen created by combined chemothermal electrolysis employing heat and nuclear power. Another variety is red hydrogen, which is produced by employing nuclear thermal energy in high-temperature catalytic water splitting. Conversely, white hydrogen is hydrogen that occurs naturally (Mazloomi and Gomes, 2012).

Green hydrogen, an increasingly promising technology, has garnered momentum as a potential solution for addressing the challenges associated with transitioning to a sustainable energy future (Hassan et al., 2023). It is produced via electrolysis, a process in which water molecules are separated into their constituent hydrogen and oxygen atoms using electricity. Green hydrogen is the resultant hydrogen that comes from using sustainable energy sources like solar or wind power. The concept revolves around the production of hydrogen gas through electrolysis, employing renewable energy sources like solar, wind, or hydroelectric power. The green hydrogen pathway holds great promise for decarbonizing several industries and sectors, including electricity production, transportation, and industry. This would help global initiatives to reduce GHG emissions and mitigate the consequences of climate change (Hassan et al., 2024).

1.4 APPLICATIONS OF HYDROGEN—AN ENERGY CARRIER

Hydrogen applications are manifold and include space exploration, manufacturing, energy production, and transportation. In addition, hydrogen blending is a crucial technique for precisely blending hydrogen with other gases to raise energy efficiency and lower GHG emissions in a range of sectors, such as manufacturing and transportation. The Energy Information Administration (EIA) states that almost all hydrogen used in the US is used for food preparation, metal treatment, fertilizer manufacture, and petroleum refining (EIA, 2022). Moreover, hydrogen is essential for lowering the sulfur content of fuels in petroleum refineries. Notably, NASA was one of the first organizations to use hydrogen fuel cells to power spacecraft electrical systems and used liquid hydrogen as rocket fuel in the 1950s. Hydrogen, which comes from a variety of energy sources, has a wide range of uses in industries, transportation, energy, and other fields. A demonstration of the use of green hydrogen is shown in Figure 1.2.

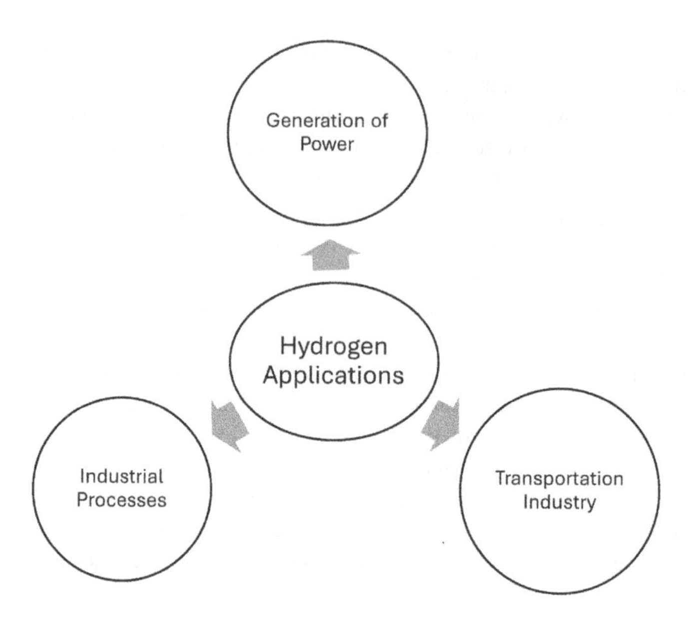

FIGURE 1.2 Different uses of hydrogen.

1.4.1 Utilization of Hydrogen in Generating Power

To meet the growing global demand for energy while reducing greenhouse gas emissions, cleaner power generation technologies are being developed as a result of the energy crisis, which is made worse by a lack of fossil fuels, the release of GHGs from conventional fossil fuel-based power generation, and rising energy consumption (Egeland-Eriksen et al., 2021). It is well known that hydrogen plays a key role in the energy transition that is required to combat climate change and decarbonize the electricity-producing industry. Furthermore, hydrogen can serve as an energy storage medium by utilizing surplus energy obtained from intermittent renewable sources (Norouzi, 2021). Hydrogen is known for being a clean energy carrier, depending on how it is produced. As a fuel, it has a high energy density and an excellent conversion efficiency. The increasing popularity of hydrogen energy is a symptom of advancements in gas turbines and fuel cell technologies that use hydrogen as their main fuel source.

1.4.1.1 Fuel Cell

Fuel cells and electrochemical devices convert the chemical energy of fuel directly into electrical energy without the need for combustion. The most common fuel used in fuel cells is hydrogen. Depending on the operational conditions and the electrolyte material utilized, these cells can be classified into multiple categories. As an example, polymer membranes such as Nafion are used as the electrolyte in proton exchange membrane fuel cells (PEMFCs). Their short start-up speeds and ability to operate at relatively low temperatures (between 50°C and 100°C) make them appropriate for a wide range of applications. On the other hand, because solid oxide fuel cells (SOFCs) can operate at temperatures between 600°C and 1000°C, they are a good option for combined heat and power (CHP) systems, auxiliary power units (APUs), and large-scale stationary power generation (Lan et al., 2023). Liquid alkaline solutions, such as KOH, are used as the electrolyte in alkaline fuel cells (AFCs), which operate at mild temperatures between 60°C and 250°C and are useful in spaceships, submarines, and small-scale stationary power generation. Molten carbonate fuel cells (MCFCs) are ideal for large-scale stationary power production and CHP systems. They work at high temperatures between 600ºC and 700ºC and use a molten carbonate salt combination such as potassium and lithium carbonates as the electrolyte. Liquid phosphoric acid serves as the electrolyte in phosphoric acid fuel cells (PAFCs), which operate at temperatures between 150ºC and 220ºC. These fuel cells are used in CHP and stationary power production systems. Protonic ceramic fuel cells (PCFCs) are used in stationary power generation and CHP systems. They function at quite high temperatures, between 500ºC and 700ºC and use a proton-conducting ceramic substance as the electrolyte (Ghorbani et al., 2021). These fuel cells are well-suited for applications like wastewater treatment, remote sensing, and small-scale power generation. There are several ways to use hydrogen in power generation, transportation, and other uses, thanks to these different kinds of fuel cells. Since fuel cells directly transform chemical energy into electrical energy, they often have higher energy efficiency than traditional combustion-based power production methods. Fuel cells that use hydrogen as their fuel only produce heat and water as byproducts, which lowers air pollution and GHG emissions (Zhang et al., 2024).

1.4.2 Employment of Hydrogen in the Transportation Industry

With 25% of worldwide emissions, the transportation sector is the second-largest source of CO_2 emissions, behind the production of power and heat. However, lowering CO_2 emissions from transportation has proven to be difficult, partly because conventional fossil fuels are more widely available and have advantages over other fuels in terms of pricing, energy density, storage, and transportation. Consequently, the main focus of attempts to reduce emissions has been on improving vehicle systems through electrification and sophisticated combustion techniques (Ghorbani et al., 2021).

1.4.2.1 Fuel Cell Electric Vehicles (FCEVs)

PEMFC is the primary power supply used by fuel cell electric vehicles (FCEVs), which offer high power output, dependable cold start capabilities, and remarkable efficiency. Unlike conventional cars with internal combustion engines, FCEVs work similarly to battery electric vehicles (BEVs) but with an electric motor and fuel cell technology. FCEVs, with their amazing 800-km driving range and 5-minute refueling time, are like conventional vehicles in that they have shorter refueling times and longer driving ranges than BEVs. Nevertheless, with FCEVs costing almost twice as much as BEVs, issues like high prices and limited infrastructure have been identified as the main concerns. Even though they are now more expensive than Internal Combustion Engine Vehicles (ICEVs) and Battery Electric Vehicles (BEVs), estimates indicate possible cost savings by 2030 (Brandon and Kurban, 2017). In addition, one major

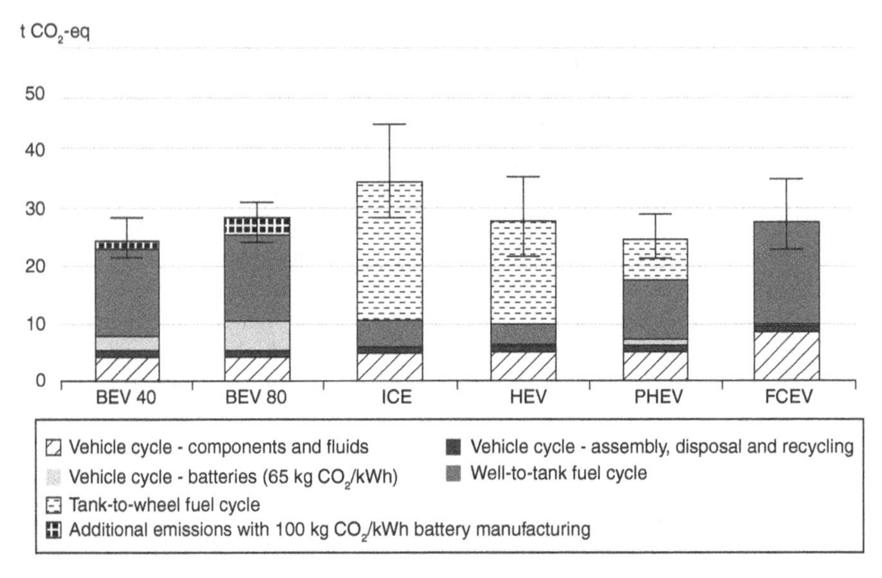

FIGURE 1.3 The powertrain of an average mid-size vehicle typically has a lifespan of 10 years (Hwang et al., 2023).

barrier is the lack of hydrogen filling stations; the US, Germany, and Japan have the fewest hydrogen filling stations in comparison to other countries. The efficiency of FCEVs in cutting carbon emissions in the transportation sector is largely dependent on how hydrogen is produced. An analysis of the life cycle of GHG emissions (Figure 1.3) of an average mid-size vehicle over 10 years, considering different powertrains, shows that plug-in hybrid electric vehicles (PHEVs), which combine electric motor and internal combustion engine systems, have lower CO_2 emissions than FCEVs (Pollet et al., 2012).

Fuel cell applications demonstrate considerable promise for broader utilization, particularly in contexts such as trucks, where high energy demands are prevalent. While BEVs may suffice for light goods and short distances, it is acknowledged that long-haul and heavy-duty transportation needs are better suited for FCEVs. Notably, in less than a year of service, XCIENT trucks in Switzerland have operated over a million kilometers. According to calculations, this fleet has cut its CO_2 emissions by 630 tons when compared to the case of regular diesel trucks. Twenty-five distinct enterprises involved in logistics, distribution, and grocery fulfillment have hired 46 XCIENT trucks since their October 2020 deployment in Switzerland (Pollet et al., 2012; Reitz et al., 2020).

1.4.3 Hydrogen Uses in Industrial Processes

Hydrogen has several applications in a wide range of industries and processes. First of all, it functions as a reactant in hydrogenation reactions, which use hydrogen atoms to break down hydrocarbons, generate molecules with a lower molecular weight, saturate compounds, and remove compounds containing sulfur and nitrogen.

Secondly, it serves as an oxygen scavenger, chemically removing trace amounts of oxygen to prevent oxidation and corrosion. Additionally, hydrogen finds application as a fuel in rocket engines, providing the necessary propulsion for space exploration. Lastly, it serves as a coolant in electrical generators, leveraging its unique physical properties to efficiently dissipate heat and maintain optimal operating conditions (Ramachandran and Menon, 1998). Most industrial applications of hydrogen involve hydrogenation, where it either saturates molecules with hydrogen atoms or breaks them apart to remove impurities like sulfur and nitrogen. These reactions often require high-pressure, high-purity hydrogen. The chemical and petroleum industries are the biggest consumers of hydrogen. Ammonia production takes the lead, using nearly half of all industrial hydrogen. Petroleum processing follows at around 37%, with methanol production at 8%. Environmental regulations are expected to further increase hydrogen usage in petroleum processing.

Hydrogen's Role in Petroleum Refining:

Hydrocracking: This process combines cracking heavier hydrocarbons with hydrogenation to create lighter, more valuable fuels.
Hydro processing: Hydrogen is used in various refining processes to improve product quality and remove impurities.

Creating Petrochemicals with Hydrogen:

Methanol: Methanol is the primary petrochemical made with hydrogen; it's produced by reacting hydrogen and carbon monoxide under high pressure and temperature.
Other Applications: Hydrogen is also used to create various other petrochemicals, such as butyraldehyde, acetic acid, and many more.

Beyond Refining:

Plastics Recycling: A newer application uses hydrogenation to break down recycled plastics into reusable lighter molecules.
Fertilizer Production: Hydrogen is crucial for ammonia production, the backbone of the fertilizer industry.
Nickel Production: The Sherritt-Gordon process utilizes hydrogen to extract nickel from sulfate solutions.

1.5 HYDROGEN SEPARATION

With predictions that global energy consumption will nearly double by 2050 and increasing pressure from the political, economic, and environmental domains on the remaining fossil fuel reserves, it is critical to resolve the many scientific and technological obstacles preventing the shift to a sustainable and competitive hydrogen economy (Adhikari and Fernando, 2006). In response to these demands, the President initiated the $1.2 billion Hydrogen Fuel Initiative in January 2003, which is a coordinated attempt to improve the safe and profitable production and storage of hydrogen. However, large-scale hydrogen production in a variety of industries frequently requires costly separation and purification procedures, which drives up the cost of hydrogen. Whatever the process used to produce hydrogen, there will always be a need for an economical and effective way to separate hydrogen from less desirable species. Currently, fractional/cryogenic distillation, membrane separation, or pressure swing adsorption (PSA) are the three main methods used in hydrogen purification. Although fractional/cryogenic and PSA distillation systems are functional, they are usually considered to be energy- and cost-intensive methods for purifying and separating hydrogen. Moreover, neither technique achieves the necessary purity levels for the intended uses in the hydrogen economy. On the other hand, membrane separation stands out as the most promising technique because of its low energy usage, ability to operate continuously, drastically lower investment costs, ease of operation, and ultimately, cost-effectiveness (Bredesen et al., 2004).

Even though PSA and cryogenic distillation technologies are widely used in industry, they often require multiple units along with supplementary wash columns for CO and CO_2 removal (Wu et al., 2012). Pressure-driven membrane processes are favored for hydrogen production due to their lower energy consumption and ability to produce high-purity hydrogen. Electrochemical membrane technology utilizing Proton Exchange Membranes (PEMs) has emerged as an attractive option for hydrogen separation, proving effective in recovering hydrogen from various hydrogen-rich gas mixtures like H_2/N_2, $H_2/CO_2/CO$, and H_2/CH_4 (Gardner and Ternan, 2007). EHS offers numerous advantages over other methods, such as single-step high-purity hydrogen production, the potential for simultaneous compression and separation of hydrogen, and achieving separation at low cell voltages with high efficiency. Moreover, this electrochemical approach can concentrate CO_2 from the feed stream without the need for extra processing steps, thereby aiding in GHG emission reduction (Bouwman, 2015).

1.5.1 Cryogenic Distillation Technologies

Many different sectors use cryogenic distillation techniques to purify industrial gases such as hydrogen, oxygen, nitrogen, argon, helium, and natural gas (Agrawal et al., 2000). A cryogenic technique is employed to produce high-purity nitrogen, where crucial process design parameters are established. This pure nitrogen is essential for applications like neutrino detection in a scintillation detector, which necessitates a working temperature of 77 K when using activated carbon (Peng et al., 2018). Bio-methane is subjected to a cryogenic process for removing CO_2 and other impurities from it as detailed by Baena-Moreno et al. (2019). The advantages and limitations of cryogenic methods are highlighted through the achievement of high-purity products and the associated high energy costs, respectively. One important area that addresses both opportunities and limitations in this purifying technique is the cryogenic CO_2 collection method (Song et al., 2019), including the provision of cold energy, high operational costs, and impurity concerns in the final product. Notably, the cryogenic approach to CO_2 separation avoids issues commonly associated with chemical

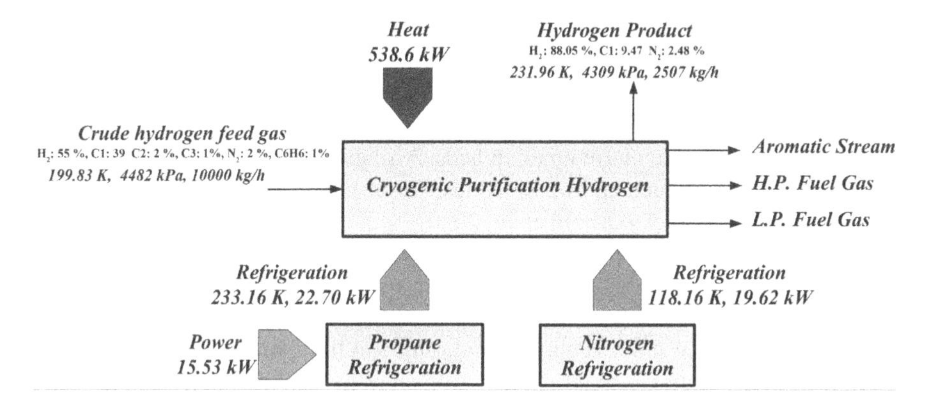

FIGURE 1.4 Diagram illustrating the entire system for cryogenic purification of gas supply (Aasadnia et al., 2021).

solvents or physical sorbents. Impurities in helium are effectively removed via a cryogenic adsorption process, with results indicating that the use of a vacuumed bed during regeneration is the optimal method for increasing helium purity. CO_2 separation from flue gases is achieved through a cryogenic-based process, leading to significant energy consumption reduction when integrating the capture process with cryogenic nitrogen purification systems. Additionally, CO_2 elimination from natural gas during liquefied natural gas (LNG) production is optimized to reduce energy consumption (Pellegrini et al., 2019). Biogas purification technologies are assessed with a focus on CO_2 removal, with cryogenic systems demonstrating minimal bio-methane losses compared to other methods. An efficient cryogenic system is proposed for removing nitrogen from natural gas (Hamedi et al., 2018), indicating the single-column system's preference over the multi-column process due to lower component numbers, reduced maintenance costs, and simpler operation.

As shown in Figure 1.4, Majid et al. presented a cryogenic module intended to extract hydrogen from a feed gas mixture comprising nitrogen and linear, branched, or cyclic hydrocarbons. The module takes a hydrogen-enriched stream from the cryogenic separation heat exchangers while the hydrogen-lean stream is processed in the cryogenic sections to remove non-hydrogen components and generate aromatic, low-pressure, and high-pressure fuel streams. The findings indicate that heat exchangers are primarily responsible for over 88.4% of energy destruction, with valves playing a much smaller role in this process (Aasadnia et al., 2021).

1.5.2 Pressure Swing Adsorption (PSA)

Pressure swing adsorption has quickly become a highly effective method for gas separation with active applications in air drying, hydrogen purification, landfill gas separation, branched and linear paraffin separation, small-to medium-scale air fractionation, and more recently, carbon capture and biogas upgrade (Liu et al., 2020). In 1966, a steam reformer plant in Toronto and the first commercial H_2 PSA unit, which consisted of four adsorbers and a tail gas drum, were put into service. Large-scale multi-bed processes began to be developed in the late 1970s; the first unit, a 10-bed, 42 MMSCFD machine was deployed at the German refinery Wintershall Lingen in 1977. There are currently more than 1000 Polybed H_2 PSA systems in use throughout the world, some of which have up to 16 beds. A single train of a modern Polybed H_2 PSA system may produce up to 240 MMSCFD of ultra-pure hydrogen, handling a wide variety of feedstock (Grimm et al., 2020).

In addition to producing extraordinarily high-purity hydrogen at values between 99 and 99.999+%, PSA systems can provide significant cost advantages over scrubbing systems. PSA units also have the advantage of being simple to operate because they don't require circulating solutions or revolving machinery. Moreover, PSAs improve plant energy efficiency by supplying 60%–90% of the total heat input by fueling steam reformer burners with low-pressure tail gas. Common adsorbents in industrial H_2 PSA units do not experience deactivation phenomena common to their counterparts used in processes like isomerization, cracking, and hydrotreating; this is because PSA cycles generally run at ambient temperatures, unlike catalysts used in refining operations. The processes of coking and sintering usually take place at temperatures higher than 400°C–500°C, or above 100°C when olefins are present. The available feed pressure or the necessary product pressure is used to determine the adsorption pressure. Most feed streams fall between 10 and 40 bar; however, there may be outliers that are as low as 5 bar or as high as 65 bar. Desorption pressure is usually set at a pressure that is 1–3 bar above atmospheric pressure. When an H_2 PSA bed is densely packed with adsorbent particles that have a diameter of 1–4 mm, the pressure drop can be as high as 1 bar (Liu et al., 2020).

Majlan et al. (2009) used activated carbon as the adsorbent material in their investigation and a compact pressure swing adsorption (CPSA) method to recover pure hydrogen from a combination of $H_2/CO/CO_2$. The CPSA system is made up of four adsorption beds that work simultaneously in different phases of the pressure swing adsorption (PSA) process cycle. Pressurization, adsorption I (feeding from the prior bed), adsorption II (direct feed), blowdown, and purging are the five essential processes in the CPSA operation. In contrast, the PSA steps in the cycle are performed simultaneously, whereas in a standard PSA arrangement, they are performed sporadically on alternate adsorbers. The feed for the second bed in the CPSA system is obtained from the first bed's output during the adsorption phase. The mixes of hydrogen and carbon dioxide move through two beds in the CPSA one after the other before leaving the system. Purified hydrogen can be the ultimate product released from the CPSA because CO is adsorbed in both the first and second beds (Majlan et al., 2009). As shown in Figure 1.5, the process cycle control system and valves were made to support both manual and automatic operations.

He stated that the CPSA achieved the CO_2 adsorption capacity of 2.05 mmol / g and the CO adsorption capacity (STP) of 0.55 mmol/g . In 60 cycles lasting 3600 seconds, the CPSA was able to reduce the CO concentration in an $H_2/CO/CO_2$ mixture from 4000 ppm to 1.4 ppm and the CO_2 concentration from 5% to 7.0 ppm CO_2. The resultant H_2 purity was 99.999% with a product throughput of 0.04 kg H_2/kg adsorbent, a purge/feed ratio of 0.001, and a vent loss/feed ratio of 0.02 (Majlan et al., 2009).

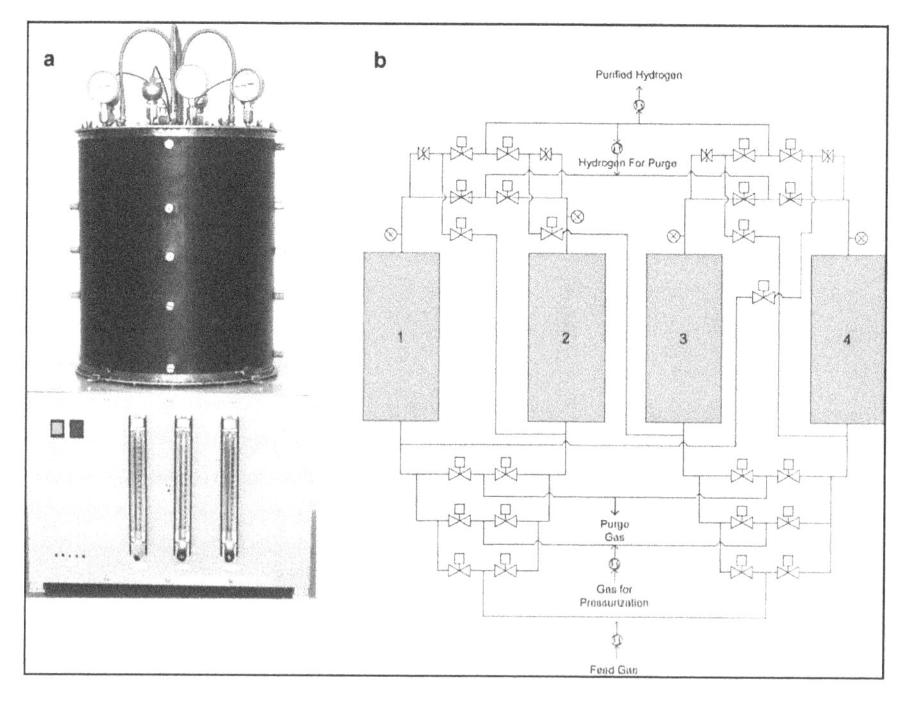

FIGURE 1.5 (a) CPSA system, (b) schematic CPSA (Majlan et al., 2009).

1.5.3 Technology of Membrane Separation

Researchers are very interested in the membrane separation approach (Figure 1.6), which has a flexible structure, high energy efficiency, and compatibility with hydrogen (Lider et al., 2023).

Conversely, low-purity or low-pressure hydrogen streams can be recovered using sophisticated membrane technology, even in situations where PSA or cryogenic methods might not be financially feasible. However, there are certain drawbacks to membrane technology, including a lowered overall recovery rate of 85%–90%, a middling purity hydrogen output of 90%–95%, and a lower feed flow. The part of the raw material that makes it through the membrane is referred to as "permeate"; the remaining fraction is referred to as "retentate". Two fundamental concepts that are directly related to membrane separation are permeability and selectivity. The selectivity and separation of the membrane provide a quantitative assessment of the distribution ratio of various components between the retentate and permeate, as well as the membrane's ability to release the required component. Since hydrogen passes through thinner membranes more quickly, improving the mechanical characteristics and hydrogen permeability of membranes can hasten the process of hydrogen permeation. Strong mechanical integrity guards against thin membrane distortion and damage brought on by the pressure differential between the sidewalls of the inlet and outflow during hydrogen penetration (Cardoso et al., 2018).

Excellent chemical, mechanical, and thermal stability, high hydrogen permeability, affordable preparation costs, little energy consumption during preparation,

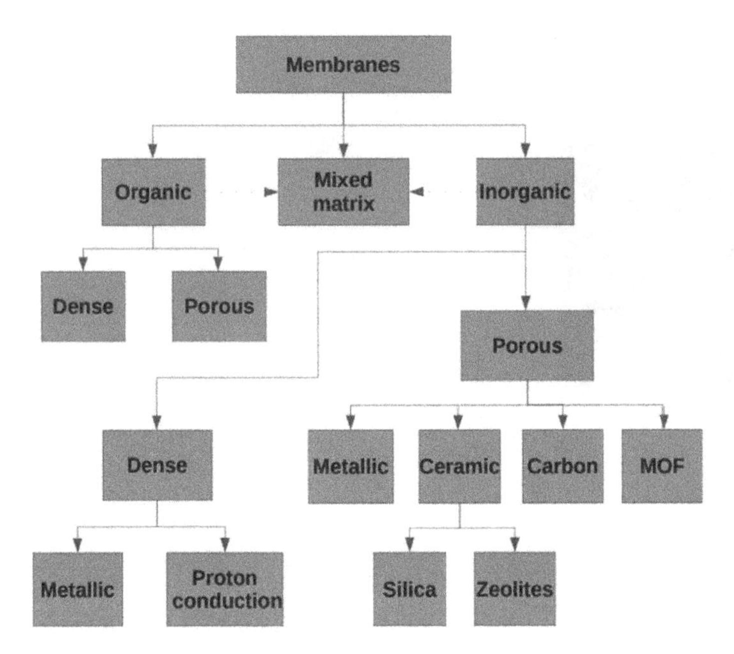

FIGURE 1.6 Suggested classification of membranes (Lider et al., 2023).

and extended service life are all desirable qualities in a membrane for hydrogen separation. Membranes for H_2 separation fall into three categories: inorganic, organic (polymeric), and mixed matrix membranes (MMMs) depending on the components that make them up. Hydrogen separation makes use of both dense and porous membranes. Because of their high permeability and affordability, porous membranes made of metals, polymers, or ceramics are a good choice for jobs involving the separation of hydrogen at low temperatures and pressures. Conversely, dense metal membranes like palladium are the preferred material for applications requiring greater selectivity in hydrogen purification as well as tasks involving high-pressure and high-temperature hydrogen separation because of their high selectivity for hydrogen. However, their production costs may be higher, and their chemical and thermal stability may be lower (Kamaroddin et al., 2018).

1.6 CONCLUSION

Hydrogen, the most abundant element in the universe, is a clean and versatile energy carrier that can transform sectors like transportation and industrial processes. Advanced separation technologies like pressure swing adsorption, membrane separation, and cryogenic distillation are crucial for harnessing its potential. However, these techniques face challenges like efficiency, cost, and scalability. Innovations in material science and engineering are needed to improve the purity and economic viability of hydrogen production. As research progresses, more robust and efficient separation methods will enable large-scale hydrogen deployment. The future of hydrogen energy is promising, with the potential to reduce GHG emissions and foster energy independence. Continued investment in research, supportive policies, and infrastructure will help realize the full benefits of hydrogen as a sustainable energy system.

REFERENCES

Aasadnia, M., Mehrpooya, M., Ghorbani, B., 2021. A novel integrated structure for hydrogen purification using the cryogenic method. *J Clean Prod* 278, 123872.

Adhikari, S., Fernando, S., 2006. Hydrogen membrane separation techniques. *Ind Eng Chem Res* 45, 875–881.

Agrawal, R., Herron, D.M., Rowles, H.C., Kinard, G.E., 2000. Cryogenic technology. In *Kirk-Othmer Encyclopedia of Chemical Technology*. John Wiley & Sons, pp. 40–65.

Arcos, J.M.M., Santos, D.M.F., 2023. The hydrogen color spectrum: techno-economic analysis of the available technologies for hydrogen production. *Gases* 3, 25–46.

Baena-Moreno, F.M., Rodríguez-Galán, M., Vega, F., Vilches, L.F., Navarrete, B., Zhang, Z., 2019. Biogas upgrading by cryogenic techniques. *Environ Chem Lett* 17, 1251–1261.

Bouwman, P., 2015. Fundamentals of electrochemical hydrogen compression. In *PEM Electrolysis for Hydrogen Production—Principles and Applications*; Bessarabov, D., Wang, H., Li, H., Zhao, N., Eds. CRC Press, p. 269.

Brandon, N.P., Kurban, Z., 2017. Clean energy and the hydrogen economy. *Phil Trans R Soc A* 375, 20160400.

Bredesen, R., Jordal, K., Bolland, O., 2004. High-temperature membranes in power generation with CO_2 capture. *Chem Eng Process Process Intensif* 43, 1129–1158.

Cardoso, S.P., Azenha, I.S., Lin, Z., Portugal, I., Rodrigues, A.E., Silva, C.M., 2018. Inorganic membranes for hydrogen separation. *Sep Purif Rev* 47, 229–266.

Egeland-Eriksen, T., Hajizadeh, A., Sartori, S., 2021. Hydrogen-based systems for integration of renewable energy in power systems: achievements and perspectives. *Int J Hydrogen Energy* 46, 31963–31983.

Energy Information Administration (EIA), 2022. *Hydrogen Explained. Production of Hydrogen*. EIA.

Gardner, C.L., Ternan, M., 2007. Electrochemical separation of hydrogen from reformate using PEM fuel cell technology. *J Power Sources* 171, 835–841.

Ghorbani, B., Rahnavard, Z., Ahmadi, M.H., Jouybari, A.K., 2021. An innovative hybrid structure of solar PV-driven air separation unit, molten carbonate fuel cell, and absorption--compression refrigeration system (Process development and exergy analysis). *Energy Rep* 7, 8960–8972.

Grid, N., 2023. The hydrogen colour spectrum. *J Hydrogen Energy*, 48(12), 5505–5515.

Grimm, A., de Jong, W.A., Kramer, G.J., 2020. Renewable hydrogen production: a techno-economic comparison of photoelectrochemical cells and photovoltaic-electrolysis. *Int J Hydrogen Energy* 45, 22545–22555.

Hamedi, H., Karimi, I.A., Gundersen, T., 2018. Optimal cryogenic processes for nitrogen rejection from natural gas. *Comput Chem Eng* 112, 101–111.

Hassan, Q., Abdulateef, A.M., Hafedh, S.A., Al-samari, A., Abdulateef, J., Sameen, A.Z., Salman, H.M., Al-Jiboory, A.K., Wieteska, S., Jaszczur, M., 2023. Renewable energy-to-green hydrogen: a review of main resources routes, processes and evaluation. *Int J Hydrogen Energy* 46, 17383–17408.

Hassan, Q., Hafedh, S.A., Mohammed, H.B., Abdulrahman, I.S., Salman, H.M., Jaszczur, M., 2024. A review of hydrogen production from bio-energy, technologies and assessments. *Energy Harvesting Syst* 11, 20220117.

Hwang, J., Maharjan, K., Cho, H., 2023. A review of hydrogen utilization in power generation and transportation sectors: achievements and future challenges. *Int J Hydrogen Energy* 48, 28629–28648.

Ishida, K., Kainuma, R., Ohnuma, I., Omori, T., Takaku, Y., Hagisawa, T., 2017. Hydrogen gas generating member and hydrogen gas producing method therefor. Patent number US9617622B2.

Kamaroddin, M.F.A., Sabli, N., Abdullah, T.A.T., 2018. Hydrogen production by membrane water splitting technologies. In *Advances in Hydrogen Generation Technologies*; Eyvaz, M., Ed. IntechOpen, pp. 19–37.

Lan, Y., Lu, J., Mu, L., Wang, S., Zhai, H., 2023. Waste heat recovery from exhausted gas of a proton exchange membrane fuel cell to produce hydrogen using thermoelectric generator. *Appl Energy* 334, 120687.

Lider, A., Kudiiarov, V., Kurdyumov, N., Lyu, J., Koptsev, M., Travitzky, N., Hotza, D., 2023. Materials and techniques for hydrogen separation from methane-containing gas mixtures. *Int J Hydrogen Energy* 48, 28390–28411.

Liu, B., Yu, X., Shi, W., Shen, Y., Zhang, D., Tang, Z., 2020. Two-stage VSA/PSA for capturing carbon dioxide (CO_2) and producing hydrogen (H_2) from steam-methane reforming gas. *Int J Hydrogen Energy* 45, 24870–24882.

Majlan, E.H., Daud, W.R.W., Iyuke, S.E., Mohamad, A.B., Kadhum, A.A.H., Mohammad, A.W., Takriff, M.S., Bahaman, N., 2009. Hydrogen purification using compact pressure swing adsorption system for fuel cell. *Int J Hydrogen Energy* 34, 2771–2777.

Mazloomi, K., Gomes, C., 2012. Hydrogen as an energy carrier: prospects and challenges. *Renewable Sustainable Energy Rev* 16, 3024–3033.

McCay, M.H., Shafiee, S., 2020. Hydrogen: an energy carrier. In *Future Energy: Improved, Sustainable and Clean Options for Our Planet*; Tyagi, V. K., Alahmer, A. A. Z., Al-Khulaifi, M. S. N. Eds. Elsevier, pp. 475–493. https://doi.org/10.1016/B978-0-08-102886-5.00022-0

Momirlan, M., Veziroglu, T.N., 2005. The properties of hydrogen as fuel tomorrow in sustainable energy system for a cleaner planet. *Int J Hydrogen Energy* 30, 795–802.

Norouzi, N., 2021. Assessment of technological path of hydrogen energy industry development: a review. *Iranica J Energy Environ* 12, 273–284.

Pellegrini, L.A., De Guido, G., Valentina, V., 2019. Energy and exergy analysis of acid gas removal processes in the LNG production chain. *J Nat Gas Sci Eng* 61, 303–319.

Peng, Z., Zhongjun, H., Bingming, W., Qing, L., 2018. Design of a nitrogen purification system with cryogenic method for neutrino detection. *Appl Radiat Isot* 137, 194–198.

Pollet, B.G., Staffell, I., Shang, J.L., 2012. Current status of hybrid, battery and fuel cell electric vehicles: from electrochemistry to market prospects. *Electrochim Acta* 84, 235–249.

Ramachandran, R., Menon, R.K., 1998. An overview of industrial uses of hydrogen. *Int J Hydrogen Energy* 23, 593–598.

Reitz, R.D., Ogawa, H., Payri, R., Fansler, T., Kokjohn, S., Moriyoshi, Y., Agarwal, A.K., Arcoumanis, D., Assanis, D., Bae, C., others, 2020. IJER editorial: the future of the internal combustion engine. *Int J Engine Res* 1, 3–10.

Serpone, N., Lawless, D., Terzian, R., 1992. Solar fuels: status and perspectives. *Sol Energy* 49, 221–234.

Song, C., Liu, Q., Deng, S., Li, H., Kitamura, Y., 2019. Cryogenic-based CO_2 capture technologies: state-of-the-art developments and current challenges. *Renewable Sustainable Energy Rev* 101, 265–278.

Winter, I.C.-J., 1999. From fossil fuels to energies-of-light. *J Hydrogen Energy Syst Soc Jpn* 24, 1–2.

Wu, X., Benziger, J., He, G., 2012. Comparison of Pt and Pd catalysts for hydrogen pump separation from reformate. *J Power Sources* 218, 424–434.

Zhang, L., Jia, C., Bai, F., Wang, W., An, S., Zhao, K., Li, Z., Li, J., Sun, H., 2024. A comprehensive review of the promising clean energy carrier: hydrogen production, transportation, storage, and utilization (HPTSU) technologies. *Fuel* 355, 129455.

2 Different Types of Hydrogen Membranes and Its Materials

2.1 CLASSIFICATION OF HYDROGEN MEMBRANES

Hydrogen can be extracted from gaseous mixtures that are created via a variety of chemical reactions or made by removing other components. Crude hydrogen-based gases derived from steam reforming or partial oxidation of hydrocarbons include carbon dioxide, water vapor, carbon monoxide, and methane among other co-products, byproducts, and residual reactants. Hydrogen produced by water electrolysis includes moisture and some oxygen. Hydrocracking, hydrotreating, and brine electrolysis all yield byproducts or unreacted hydrogen to differing degrees of purity. Hydrogen of moderate or high purity is produced from these hydrogen-rich gases via preparatory separation and purification activities. The term "separation" refers to all first-stage hydrogen concentration processes, whereas "purification" refers to later procedures that enhance produced hydrogen. Several methods for separating and purifying hydrogen are known.

The chemistry, structure, and many uses for hydrogen membranes determine their classification. One may categorize membranes as synthetic or natural. Further classification of synthetic membranes is made into three groups: mixed matrix membranes (MMMs), inorganic membranes, and organic membranes (polymer membranes). Though organic membranes have poor thermal stability, synthetic or natural polymer membranes are currently mostly used in industrial gas generation because of their low cost, scalability, and outstanding resistance to high-pressure drop. Usually, its range of separation capacity is 363–373 K (Jokar et al., 2023). In the meanwhile, although being more expensive, inorganic membranes offer the benefits of good thermal, chemical, and mechanical stabilities with minimal plasticization and a regulated pore-size distribution for better control over selectivity and permeability. Dense (nonporous) and porous membranes are the two morphological categories into which membranes fall. Proton-conducting membranes, whether metallic or ceramic, are dense inorganic materials. Whereas porous membranes (silica, zeolite, and carbon molecular sieve (CMS)) can separate molecules according to their sizes, morphologies, and affinity for permeable membranes, dense membranes can separate molecules according to their solubility and diffusivity. Among the five separation techniques—Knudsen diffusion, surface diffusion, capillary condensation, molecular screening, and solution-diffusion—one or more can be responsible for

DOI: 10.1201/9781003590682-2

H_2 separation over a membrane (Chen et al., 2024). Eventually, the contribution of different mechanisms to specific materials affects their overall performance and separation efficiency. Porous membranes have a diffusion mechanism that may include Knudsen diffusion, surface diffusion, capillary condensation, or molecular sieving and is largely governed by membrane morphology (pore size) and size of diffusion molecules. Smaller holes in Knudsen diffusion cause the material of diffusion to collide more with the pore wall. A material moves along the surface and becomes adsorbed on the pore wall's surface during surface diffusion. The compounds would move under pressure from a high-concentration to a low-concentration zone. Size of the pores and interactions between the penetrant and the pore wall are the causes of capillary condensation. Capillary pressure controls the transmembrane diffusion rate and improves separation efficiency, therefore determining the driving force of mass transport. Molecule sieving takes place when the pore diameter gets close to the size of the diffusion molecule. Consequently, molecule size largely determines the activation energy that must be overcome before diffusion may start. Diffusion of solutions governs mass transport in a dense, nonporous membrane. Owing to the chemical potential gradient, permeability molecules dissolve in the membrane matrix and diffuse through it (Chen et al., 2024).

2.2 DENSE METALLIC MEMBRANES

Dense metallic membranes are a promising method for extracting hydrogen from gas mixtures. These membranes are often composed of palladium alloys, which have a high hydrogen affinity and can separate high-purity hydrogen at high temperatures. Recent difficulties for metallic membranes include integrating high sulfur tolerances with high hydrogen permeability and low prices. Hydrogen is transported through metallic membranes by a solution-diffusion mechanism that involves molecular hydrogen dissociation and atomic hydrogen diffusion within the metal lattice. Usually employed between 300°C and 600°C, dense metallic membranes are generally used in syngas treatment for electricity generation. The most often used support materials for composite Pd membranes are porous ceramic and stainless steel, both of which exhibit remarkable chemical-thermal stability (Al-Mufachi et al., 2015). But because Pd is so expensive and scarce, its widespread use is still restricted. Less expensive substitutes, such as Nb-, Zr-, and V-based alloys as well as Ni-Nb-Zr and V-Ni amorphous alloys, are thus receiving greater attention. These metallic membranes generate extremely pure hydrogen (99.99%), but they have a bad surface quality that includes the formation of surface oxide layers. The solution-diffusion idea underpins the filtering mechanism across the membrane, according to which gaseous particles dissolve into the membrane matrix first and then pass through it at different speeds based on the partial pressure differential between the feed and permeate sides (Jokar et al., 2023). Many publications explain various strategies for fabricating such composite membranes. Sputtering, electroless deposition, CVD, and electroplating are the most frequent processes employed. These processes produce a thin layer of palladium or palladium-based alloy either inside the holes of porous supports or on their surface. Porous, smooth on the surface, highly permeable, thermally stable, and metallic sticky should

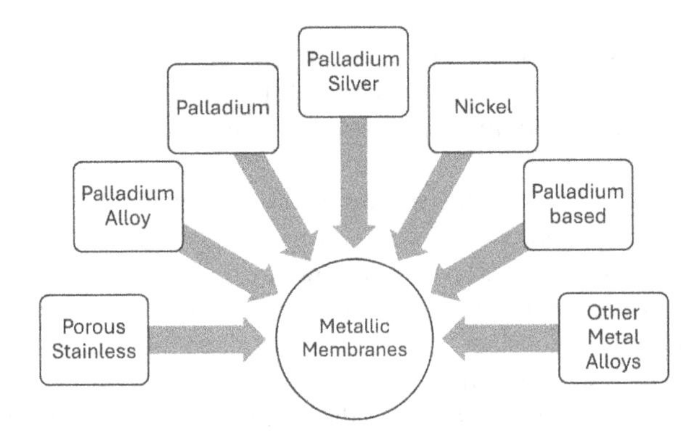

FIGURE 2.1 Different types of metallic membranes applicable in hydrogen separation and purification.

be the characteristics of high-quality metallic composite membrane support. Besides, its surface needs to be free of flaws. Moreover, the manufacturing procedure should be easy and feasible to produce a large amount of membrane. Hydrogen separation is best served by palladium (Pd) membranes, which have been studied and made commercial. Their key is palladium's strong affinity for hydrogen, which allows it to absorb and selectively pass the gas through. These membranes are normally made by depositing a thin palladium layer on a supporting ceramic or stainless steel basis. Palladium-silver (Pd-Ag) alloys are becoming increasingly popular for their superior mechanical strength and hydrogen permeability. Silver improves permeability without losing selectivity. But Pd is not confined to silver. It can be further alloyed with gold, copper, or nickel (Pd-X membranes) to improve selectivity, stability, and cost. Nickel (Ni) membranes provide a low-cost option with good hydrogen permeability (Pati et al., 2017). However, they frequently require greater operating temperatures. Pd-based composite membranes close the gap between strength and performance. These membranes contain palladium or its alloys within a ceramic matrix, resulting in high hydrogen permeability and selectivity while increasing mechanical strength and stability. Porous stainless steel membranes, with regulated porosity, can also be used. These membranes rely on hydrogen transport through the metal, which may be enhanced by surface changes or catalytic coatings. Other metal alloys, such as tantalum, titanium, and zirconium, are being studied for their potential as hydrogen-selective membranes, with a wide range of characteristics for different working situations (Figure 2.1).

2.2.1 Palladium Membrane

Palladium membranes are thin, dense, and very selective for hydrogen. Palladium membranes allow hydrogen molecules to pass through while preventing the passage of other gases, including nitrogen, oxygen, and carbon dioxide. This is because hydrogen molecules are tiny and can easily pass through the palladium lattice. Palladium membranes often have a hydrogen selectivity of more than 99.9%.

Vanadium, niobium, and tantalum are refractory metals because, at lower temperatures, their permeability is orders of magnitude greater than palladium's. Compared to palladium, these metals are more affordable and heat resistant. However, manufacturing of these metal membranes has come to a standstill due to hydrogen embrittlement, which causes them to lose their flexibility when subjected to 6.9 bar hydrogen at ambient temperature. In addition, metals limited use is due to their poor surface characteristics, which cause surface oxide layers to develop and hinder hydrogen transport. Due to their high tolerance to hydrocarbon flows, capacity to self-catalyze H_2 dissociation processes, and excellent permeability, palladium (Pd) membranes have lately attracted a great deal of interest. Aside from its catalytic aptitude for hydrogen recombination, the Pd membrane has a strong resistance to hydrogen embrittlement. Speedy hydrogen diffusion is possible due to the Pd lattice's ability to split hydrogen into monatomic hydrogen. In a volume, Pd can absorb about 600 orders of magnitude of hydrogen without changing its physical properties or structural integrity (Rahimpour et al., 2017).

Palladium and palladium-based alloy layers are, however, defined by the deposition method and the pore size of the support. According to Uemiya, the support quality—including features like a narrow pore-size distribution and the quantity of surface imperfections—strongly affected the palladium layer thickness. The relationships that were proposed between the pore size and the thickness of the Pd layer were as follows: 13 μm vs 0.3 μm, 4.5 μm versus 0.2 μm, 2.2 μm versus 0.1 μm, and 0.8 μm versus 5 nm. The researchers Mardilovich et al. showed that in order to create a dense layer by electroless plating, the palladium thickness needed to be at least three times wider than the largest pores in the support.

Matsumura and coworkers investigate the hydrogen permeation mechanism through palladium membranes using theoretical modeling and experimental validation. The study gives precise insights into the diffusion kinetics, surface reactions, and mass transport mechanisms that drive hydrogen permeation, thereby improving our fundamental understanding of palladium membrane behavior. Huo et al. present a detailed assessment of performance-enhancing options for palladium membrane reactors used in hydrogen production operations. The article discusses membrane module design, reactor layout, catalyst integration, and system optimization, providing useful insights for increasing reactor efficiency and hydrogen generation. Rahimpour et al. demonstrated the feasibility and effectiveness of using a high-flux palladium membrane for efficient hydrogen purification. The membrane exhibited high hydrogen permeability and selectivity, allowing the separation of hydrogen from complex gas mixtures found in refinery off-gas streams. They found that the high-flux palladium membrane reactor holds promise as a low-cost, energy-efficient alternative for separating hydrogen from refinery off-gas streams. The membrane-based technique has several advantages, including excellent separation efficiency, a small footprint, and the potential for integration with existing refinery equipment. Ghazi et al. developed and characterized palladium membranes for high-temperature hydrogen separation applications. They wanted to know whether palladium membranes could be used at high temperatures and how well they separated hydrogen from gas mixtures. Their studies revealed the palladium membranes ability to selectively permeate hydrogen at high temperatures. This membrane displayed strong hydrogen permeability and selectivity, indicating that it has the potential to efficiently separate hydrogen from gas streams even at high temperatures. Furthermore, palladium membranes are a viable solution for effective hydrogen separation in high-temperature processes such as steam methane reforming and biomass gasification, hence contributing to the improvement of sustainable hydrogen generation technologies.

2.2.1.1 Advantages of Palladium Membrane

1. Strong attraction to hydrogen, enabling efficient passage of hydrogen molecules through the membrane.
2. Demonstrates a high level of selectivity for hydrogen, efficiently isolating hydrogen from other gases such as nitrogen, carbon dioxide, and methane.
3. These materials exhibit chemical and thermal stability, rendering them highly resistant to harsh operating conditions and long-term degradation.
4. It can selectively allow hydrogen to pass through while blocking contaminants, resulting in the production of high-quality hydrogen with a purity level of over 99.9%.
5. Palladium membranes are dense and may be readily incorporated into current systems, offering versatility in design and space efficiency.
6. Functions at relatively low temperatures (200°C–500°C), resulting in decreased energy usage as compared to conventional separation techniques.
7. Can be expanded to larger dimensions to fulfill the demands of industrial-scale hydrogen production.
8. Can be utilized to extract hydrogen from process streams and waste gasses, enhancing overall efficiency and decreasing emissions.
9. Commonly employed in fuel cell systems to segregate hydrogen from the anode and oxygen from the cathode, guaranteeing optimal efficiency and performance.
10. Facilitates the manufacture and consumption of clean hydrogen fuel, hence contributing to the decrease of greenhouse gas emissions.

2.2.1.2 Disadvantages of Palladium Membrane

1. Palladium is a scarce and valuable metal, which leads to high costs in the production and upkeep of palladium membranes.
2. The passage of hydrogen across the membrane can result in the palladium becoming brittle, which can lead to a decrease in its mechanical strength and potential failure.
3. Although palladium membranes exhibit remarkable selectivity for hydrogen, they also allow the permeation of other gases like helium, nitrogen, and carbon monoxide, so compromising the purity of hydrogen.
4. Palladium membranes exhibit optimal performance at temperatures that are lower than 500°C. At elevated temperatures, the hydrogen diffusion across the membrane diminishes, resulting in a drop in the efficiency of separation.
5. Impurities present in the input gas, such as sulfur compounds, have the potential to contaminate the palladium surface, resulting in a decrease in its ability

to allow hydrogen to pass through and a decrease in its ability to selectively separate hydrogen from other gases.

6. The membrane surface can accumulate deposits or pollutants, which can block the pores and decrease the hydrogen permeance.
7. Under specific operational circumstances, the membrane surface may experience scaling or corrosion, which can have a detrimental impact on its performance and longevity.
8. Palladium membranes may experience degradation over time as a result of hydrogen embrittlement, oxidation, or other reasons, which can diminish their efficiency.
9. Palladium membranes exhibit a comparatively significant pressure differential, leading to elevated operational expenses in the separation procedure.
10. The manufacture of palladium membranes involves the use of specialized processes and materials, which can contribute to the intricacy and expense of the process.

2.2.2 Palladium-Based Membranes

Pd-based metallic membranes, specifically, are the most efficient technology for separating high-purity hydrogen from gas mixtures containing low concentrations of hydrogen (<30%). However, for Pd-based membranes to function effectively, they need to be operated at high temperatures (>350°C) and with a significant pressure difference across the membrane surface. Due to this rationale, membrane technology was suggested to be economically viable only for high-pressure transmission pipelines as opposed to distribution networks. Furthermore, the Pd membranes exhibit a high level of sensitivity toward gas contaminants, particularly sulfur species present in the gas stream. However, the process of alloying has been demonstrated to enhance both the flow of hydrogen and the ability to withstand surface poisoning (Ma et al., 2016).

2.2.2.1 Different Types of Palladium-Based Membranes

Palladium-based membranes of various sorts provide a variety of benefits in hydrogen separation processes. By integrating different palladium alloys or making structural adjustments, these membranes can be tailored to increase selectivity for hydrogen, resulting in a better level of purity for the separated hydrogen gas. Moreover, certain palladium alloys or arrangements can enhance the permeability of hydrogen, facilitating quicker separation and greater throughput. Certain membranes are specifically engineered to withstand high temperatures, enabling them to function more efficiently under such conditions. By enhancing membrane designs and integrating reinforcing elements, mechanical stability can be enhanced, guaranteeing durability during operational situations. Moreover, these membranes can be customized to be suitable for various process scales, providing cost-efficient solutions that achieve a balance between performance, longevity, and material expenses. Application-specific features can be tuned to enhance hydrogen purification, recovery, or generation for different uses. Finally, specific membranes made of palladium are intentionally developed to address environmental concerns by minimizing energy usage, emissions, and the utilization of dangerous substances during the separation process, so boosting sustainability. The different types of palladium-based membranes are listed below.

2.2.2.1.1 Palladium-Silver (Pd-Ag) Membranes

Pd-Ag membranes are composed of a thin layer of Pd-Ag alloy that is applied onto a porous support. The composition, thickness, and structure of the alloy can be adjusted to enhance the performance of the membrane. The enhanced hydrogen permeability of Pd-Ag membranes is attributed to the alloy's elevated hydrogen solubility and diffusivity. By incorporating silver into palladium, the hydrogen permeability is improved due to the reduction in the creation of palladium hydrides, which have the potential to obstruct hydrogen transport. Palladium-silver (Pd-Ag) membranes have exhibited exceptional efficacy in separating hydrogen. They can selectively isolate hydrogen from other gases, such as nitrogen, carbon dioxide, and methane. The hydrogen permeance and selectivity of Pd-Ag membranes are contingent upon the alloy composition, operating temperature, and pressure. Palladium-silver alloys with 23–25 atomic percent (at%) of silver exhibit exceptional chemical stability toward hydrogen. These alloys are employed in the production of commercial permeator tubes, which have a thickness ranging from 100 to 150 μm, for hydrogen purification and separation.

Hatlevik et al. produced Pd alloy composite membranes that incorporate Au and/or Ag. These membranes exhibit a greater hydrogen flux compared to pure Pd membranes, as anticipated based on previous research. Under the specified test conditions of 400°C and 1.38 pressures, the rate of pure hydrogen permeation through a $Pd_{95}Au_5$ composite membrane, which was 2.3 μm thick, was measured to be 1.01 mol/m^2 s. A similar hydrogen flux of 0.97 mol/m^2 s was recorded for a 4.6 μm thicker $Pd_{80}Ag_{20}$ composite membrane at 400°C and 1.38 bar ΔP pressure differential. The higher permeability of the Pd-Ag alloy than that of pure Pd and Pd-Au is consistent with this result. Ideal Pd-Ag membranes have H_2/N_2 separation coefficients ranging from 337 to more than 90,000. By use of chemical plating, Uemiya et al. created a Pd-Ag alloy membrane on a porous alumina tube. The film had a thickness ranging from 4.5 to 6.4 picometers (pm), which is significantly thinner than the Pd films documented in previous literature by more than ten times. Their membrane had a notable hydrogen selectivity, but it was acquired by multiple iterations of the activation process, which is a complex technique. Li et al. produced a Pd-Ag alloy membrane on a thin a-alumina hollow fiber modified with a y-alumina film by using spray pyrolysis for the first time. The membrane had a reduced silver concentration compared to the solution. Research has shown that the addition of palladium to silver improves the capacity of hydrogen to pass through, as compared to pure palladium. The higher bond distance of Ag is 0.289 nm and has been noted to cause the higher H_2 permeability in Pd–Ag membranes in a number of further studies. Furthermore, it is observed that the diffusion coefficient decreases with increasing Ag presence, whereas the solubility of H_2 follows the opposite pattern and peaks between 20% and 40% Ag content. Additional research has shown that at a temperature of 350°C, the ability of hydrogen (H_2) to pass through $Pd_{77}Ag_{23}$ alloys is approximately 1.7 times greater than that of pure Pd. $Pd_{77}Ag_{23}$ membranes are commonly selected for H_2 separation because of their elevated H_2 permeability and the membrane's ability to withstand embrittlement at temperatures below 300°C. Fernandez et al. conducted an experimental study on the separation of H_2 using Pd–Ag membranes, specifically for applications involving high temperatures. At temperatures ranging from 400°C to 673°C, they observed a remarkably high permselectivity of H_2 and N_2, exceeding 200,000. The critical temperature of the α/β-phase transformation dropped from 298°C (571 K) to ambient temperature when alloyed with 23 wt% Ag, according to a study by Timofeev et al. on Pd–Ag membranes for H_2 separation. The influence of pressure (0.2–0.5 MPa) and temperature (300°C–600°C) was studied by Pinto et al., who found that the penetration flux of H_2 increased consistently with increases in both variables. A study was carried out by Ma et al. regarding the hydrogen (H_2) solubility and diffusivity in Pd–Ag membranes. The purpose of this study was to gain a better understanding of and make enhancements to H_2 permeability. Baloyi et al. conducted a study on the long-lasting and heat resistance properties of a Pd–Ag membrane while continuously exposed to H_2 gas. The membrane had a silver (Ag) content of 23 wt% and was tested under both high and low-temperature circumstances. Based on their findings, the problems with the membrane reactor's performance were due to the interaction of H_2 with the membrane's surface, leading to a decrease in H_2 permeability. In their study, Sonwane et al. utilized density functional theory (DFT) to examine the diffusivity of hydrogen (H_2) in palladium (Pd) membranes. Based on their research, they discovered that at a temperature of 456 K, the solubility of H_2 in $Pd_{70}Ag_{30}$ membranes was determined to be ten times greater than in pure Pd.

2.2.2.1.2 Palladium-Copper (Pd-Cu) Membranes

Palladium-copper (Pd-Cu) membranes are highly promising materials for hydrogen separation, owing to their exceptional hydrogen permeability and selectivity. Palladium-copper membranes are recognized for their diminished susceptibility to becoming brittle, even when exposed to lower temperatures. Pd-Cu membranes are commonly produced using electrodeposition or physical vapor deposition methods. The membranes are composed of a slender, compact Pd-Cu alloy layer that is upheld by a permeable substrate. The Pd-Cu alloy layer functions as a barrier that selectively allows hydrogen to pass through, while the substrate offers structural support and enables the movement of gas. The Pd-Cu alloy's composition can be adjusted to maximize the qualities of the membrane. Palladium increases the ability of hydrogen to dissolve and pass through, whereas copper strengthens the durability of the membrane and decreases the expense. The separation of hydrogen in Pd-Cu membranes takes place via a solution-diffusion mechanism. Hydrogen molecules undergo dissolution into the Pd-Cu alloy layer, then proceed to permeate through the membrane, and finally desorb on the opposite side. The membrane's selectivity is a result of the varying solubilities and diffusion rates of hydrogen and other gases. Hydrogen exhibits a notable solubility and diffusion rate in Pd-Cu alloys, whereas nitrogen, oxygen, and carbon dioxide display considerably lower solubilities and diffusion rates. The variation in permeabilities enables the membrane to selectively allow the passage of hydrogen. Palladium-copper (Pd-Cu) alloys have garnered significant interest as membranes due to the cost-effectiveness of copper (Cu), improved resistance to hydrogen sulfide (H_2S) compared to pure palladium (Pd), and increased hydrogen (H_2) permeability seen in certain Pd-Cu alloys. Nevertheless, the utilization

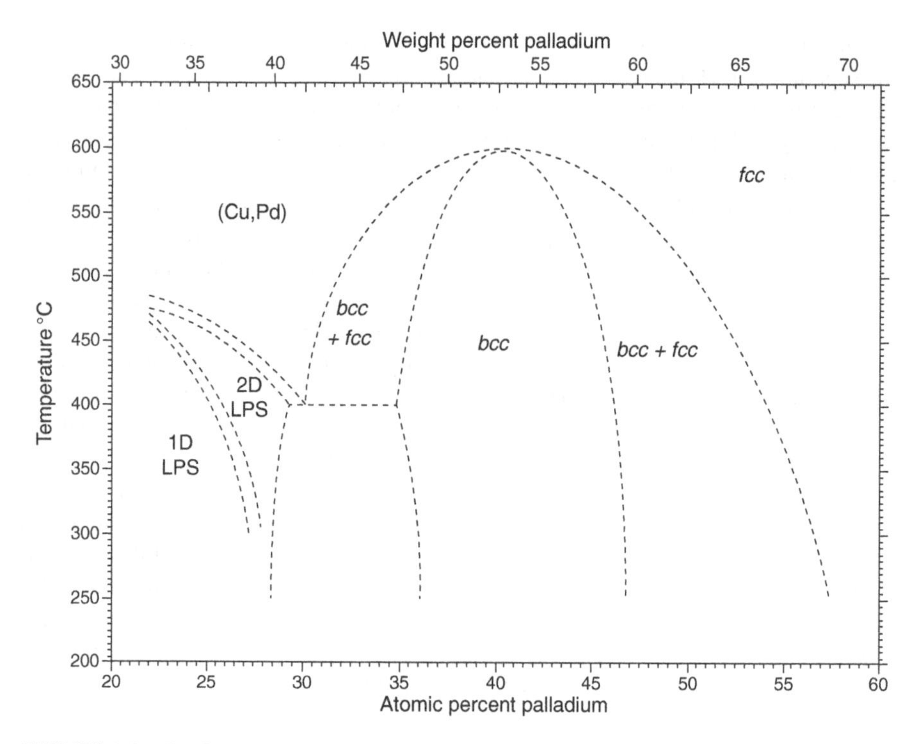

FIGURE 2.2 Pd-Cu phase diagram (Zhang and Way, 2017).

of Pd-Cu membranes for H_2 separation encounters significant obstacles primarily due to the intricate alloy phase diagram comprising multiple phase structures and the complex relationship between phase structure, composition, stability, and permeability for Pd-Cu membranes, which requires careful consideration. At temperatures beyond 600°C, Pd and Cu create a continuous face-centered cubic (fcc) solid solution. However, at 600°C, both fcc and body-centered cubic (bcc) structures can coexist, with the two phases separated by an area of mixed phases known as a miscibility gap (Figure 2.2).

In a recent study, Pomerantz et al. investigated the phase growth process of a Pd-Cu bilayer with approximately 15 wt% Cu. They determined that the creation of Pd-rich fcc alloy requires an annealing period of 225 hours at a temperature of 500°C. Furthermore, the annealing time specifically pertained to the duration it took for an fcc alloy to form on the surface. If the layer is uniform, the duration of the annealing process would need to be significantly extended. It can be quite challenging to achieve homogeneous alloying of Pd-Cu at temperatures below 450°C, even though steady-state H_2 permeation across Pd-Cu membranes can be achieved at temperatures as low as 125°C. In their study, Morreale et al. discovered that the face-centered cubic phase of Pd–Cu membranes exhibited a higher resistance to sulfur poisoning compared to the body-centered cubic phase. The researchers observed a larger level of H_2 permeation in $Pd_{80}Cu_{20}$ membranes compared to $Pd_{60}Cu_{40}$ membranes at a temperature of 560°C. However, in the absence of sulfur, the $Pd_{60}Cu_{40}$ membrane exhibited greater permeability. In a recent study, Pomerantz et al. explored the effects of H_2S on the stability and performance of Pd–Cu membranes. The membranes contained different amounts of Cu, specifically 8%, 18%, and 19%. Approximately 80% of the permeation losses were observed at a temperature of 500°C, and it was determined that these losses may be partially reversed through treatment with H_2. At a temperature of 183°C, the measured solubility of H_2 in the $Pd_{80}Cu_{20}$ membrane was approximately five times lower than that of pure Pd. Researchers conducted a study on the thermal stability of Pd-Cu and Pd-Au alloy membranes under high temperatures. At temperatures of 650°C, the Pd–Cu membranes maintained stable performance, while the Pd-Au membranes started to decline at 550°C.

2.2.2.1.3 Palladium-Gold (Pd-Au) Membranes

Palladium-Gold (Pd-Au) membranes are a type of metallic membranes that have an outstanding ability to allow hydrogen to pass through and have a high level of selectivity. Their composition consists of a thin coating of palladium that is alloyed with gold, which is applied over a porous substrate. By incorporating gold into the palladium matrix, the characteristics of the membrane are altered, resulting in enhanced hydrogen transport and decreased permeability of impurities. Within these membranes, hydrogen molecules exhibit selective diffusion through the palladium lattice, whereas other gases are predominantly excluded. This is due to hydrogen's elevated solubility and diffusivity in palladium. Incorporating gold into the palladium lattice improves the membrane's stability and durability without compromising its high hydrogen permeability. In addition, gold reduces the likelihood of palladium becoming brittle and undergoing changes in its structure, which in turn extends the lifespan of the membrane. In summary, the hydrogen separation process of Pd-Au membranes relies on the selective absorption and diffusion of hydrogen via the palladium-gold alloy. This characteristic makes them a suitable choice for a range of industrial applications that demand hydrogen with a high level of purity.

Multiple investigations have indicated potential positive outcomes for the permeability of hydrogen gas using Pd-Au alloy membranes. Chandrasekhar and Sholl conducted a computational analysis of hydrogen permeability in Pd-Au membranes with gold compositions of 4% and 15% at a pressure of 1 atmosphere. At a temperature of 327°C, the researchers observed a 5.4-fold increase in H_2 permeability in $Pd_{96}Au_4$ membranes and a 2.6-fold rise in $Pd_{85}Au_{15}$ membranes. Investigations have shown that a Pd-Au alloy membrane with a 15% Au component has higher permeability compared to pure Pd. At a temperature of 183°C, the solubility of H_2 in the $Pd_{80}Au_{20}$ membrane was found to be 12 times higher than in pure Pd. Chen et al. conducted a comparative examination on membranes made of Pd and Pd-Au (containing 8% Au). Both membranes underwent H_2S exposure, and their performance was assessed. Through the process of dissociative absorption of H_2S, the surface sites became obstructed, causing a reduction in H_2 permeation and the occurrence of bulk sulfidation. It has been observed that Pd-Au alloy membranes demonstrate superior performance compared to pure Pd, with an increased penetration rate and reduced bulk sulfidation. By subjecting the membranes to pure H_2 at a temperature of 500°C, a nearly complete regeneration was achieved. As a result, given the expensive nature of gold (Au) and its numerous performance benefits, membrane reactor designers must carefully consider the tradeoffs. McKinley patented Pd-Au in 1967 as a material for membranes used in hydrogen purification due to its superior resistance to poisoning compared to pure palladium. According to this patent, membranes containing 40 wt% of gold have been found to greatly decrease the obstruction of hydrogen permeability in gaseous streams that contain H_2S. Meanwhile, membranes containing 1–20 wt% gold demonstrated a remarkable 30% increase in hydrogen permeability. Further investigation into the permeability of hydrogen in Pd-Au alloys suggests that the most effective amount of gold to add is 11 at%. This will ensure maximum permeability at higher temperatures. This discovery is in line with the expected results from theoretical models.

Pd-Au alloys with gold contents of up to 20 at% showed improved hydrogen permeability compared to pure Pd membranes. Nevertheless, as the gold content in the alloy increases, the permeability experiences a rapid decrease. In their study, Chen and Ma investigated the resistance of a low gold content alloy ($Pd_{92}Au_8$) to H_2S at high temperatures. They observed that there was no formation of sulfide throughout the material, and the hydrogen permeation fully recovered after exposure to 55 ppm H_2S. This indicates that the decline in permeability on Pd-Au membranes was solely due to the dissociative adsorption of H_2S. Subsequent investigations conducted at colder temperatures revealed slight and lasting decreases in the hydrogen permeability during extended experiments with H_2S. With the use of sour syngas, Gade et al. investigated how alloy composition and membrane construction affected the resistance to sulfur poisoning. They found that inhibition of hydrogen permeability decreased with increasing gold content in membranes up to 20 wt%. Furthermore, the study also discovered that magnetron-sputtered membranes are more prone to poisoning than membranes produced by metallurgy. Corrosion caused the researchers to see large metal losses in sputtered alloys, which finally caused the membranes to fail early. The surface roughness of the membranes has been linked by other studies to this effect; automated membranes often have a smoother surface, which lowers H_2S absorption. Experiments revealed the ideal composition to be a gas combination with a 20 wt% gold component. For moist water gas shift (WGS) free of sulfur, permeability did not decrease. Permeability decreased by 60% nevertheless, when WGS included 20 parts per million of H_2S. In extended tests using coal-derived syngas under real industrial circumstances, the Pd-Au membranes showed remarkable resilience.

2.2.2.1.4 Palladium-Platinum (Pd-Pt) Membranes

Palladium-Platinum (Pd-Pt) membranes exhibit exceptional potential for hydrogen separation owing to their superior hydrogen permeability and selectivity. Their composition comprises a slender, compacted Pd-Pt alloy layer that is upheld by a permeable substrate. Palladium-Platinum (Pd-Pt) membranes often possess a thickness ranging from 1 to 10 μm and exhibit hydrogen permeance within the range of 10^{-6} to 10^{-4} moles per square meter per second per pascal (mol/m^2 s Pa). The hydrogen selectivity typically exceeds 100, suggesting a high level of separation efficiency. The strong hydrogen selectivity of Pd-Pt membranes is attributed to the compact structure of the alloy layer. This configuration inhibits the movement of bigger molecules, such as methane and carbon dioxide, while facilitating the permeation of hydrogen. Pd-Pt membranes demonstrate remarkable hydrogen permeability due to the ability of hydrogen atoms to dissolve into the metal lattice and diffuse through the membrane. Pt increases the hydrogen permeability by decreasing the activation energy required for hydrogen dissolution.

Platinum has been studied as an alloying element because it is more resistant to sulfur poisoning than palladium. Nevertheless, platinum's low permeability values prevent its use as a pure metal. The electroless deposition of Pd-Pt is a widely recognized technique that can be used to create thin films. Palladium selectivity can be improved and permeability can be reduced by adding between 5 and 26 wt% of Pt. Despite the unpromising permeability values, platinum significantly improves the mechanical characteristics and thermal stability of the film. This enables steady permeation at temperatures up to 650°C and reduces the occurrence of leaks during long-term operations. Additionally, the alloy exhibits exceptional resistance to contamination. According to reports, Pd_4Pt alloy membranes saw a 50% reduction in flux when exposed to 1000 ppm of H_2S. However, the membranes were able to fully restore their original flux after being cleaned. When exposed to WGS mixtures, Pd-Pt membranes also exhibited a decrease in inhibition caused by WGS, as compared to pure Pd membranes. This resulted in larger hydrogen fluxes, even though Pd-Pt membranes have lower permeability in pure hydrogen.

2.2.2.1.5 *Palladium-Nickel (Pd-Ni) Membranes*

There have been reports that nickel is an interesting alloying metal for palladium. This alloying metal results in alloys that have greater hardness and ductility, and they are also less likely to go through embrittlement. An extremely thin and dense layer of Pd-Ni alloy is used to construct Pd-Ni membranes, which are then supported on a porous substrate. It is possible to modify the alloy composition as well as the thickness to achieve optimal performance from the membrane for a particular application. The "Pd-H phase transition" is a distinctive property that is exhibited by Pd-Ni alloys. This property is characterized by the fact that the absorption of hydrogen results in a structural transformation, which in turn leads to an increase in the hydrogen permeability. Deposition of Pd-Ni membranes was accomplished through the use of electrodeposition, CVD, and ELP methods. Diffusion through grain surface rather than bulk lattice was thought to be the source of a hydrogen partial pressure exponent that is close to one. Very thin films with a hydrogen partial pressure exponent, sufficient permeability, and low selectivity were obtained via electrodeposited membranes. Selectivity increases to values greater than 4,000 when operating temperatures are increased to 600°C. It is possible that the sintering of the fine grains is the cause of this concern. Although thin films free of flaws and with reduced permeabilities have been successfully produced by CVD and ELP Pd-Ni, this alloy does not seem to be a viable option for hydrogen membranes.

2.2.2.1.6 *Palladium-Zirconium (Pd-Zr) Membranes*

Pd-Zr membranes are composed of a thin palladium-zirconium alloy layer placed over a porous support material, which is commonly ceramic or metal. The alloy layer functions as a hydrogen-selective barrier, while the support ensures mechanical stability and gas transport. The composition and thickness of the Pd-Zr alloy layer have a major impact on the membrane's performance. Higher palladium content increases hydrogen permeability, but higher zirconium content improves hydrogen selectivity. The ideal palladium-to-zirconium ratio is typically around 80:20. Nayebossadri et al. recently revealed that a minor amount of zirconium (2 at%) increased fcc-Pd-Cu's resistance to sulfur poisoning. Zr is expected to inhibit Pd and Cu surface segregation, slowing the kinetics of sulfide production. When exposed to hydrogen-containing 1000 ppm H_2S at 450°C, the $PdCu_{37}Zr_2$ ternary alloy reduces hydrogen flow by 15% less than $PdCu_{35}$.

Li et al. investigated surface modification approaches to improve the stability of Pd-Zr membranes used in hydrogen separation. They discovered that covering the membrane surface with a silica layer increased membrane stability and resistance to hydrogen embrittlement and phase changes. The study revealed the efficacy of surface modification as an approach for improving membrane integrity and extending its lifespan. Chen et al. studied the impact of zirconium content on the performance of Pd-Zr alloy membranes for hydrogen separation. They discovered that increasing the zirconium concentration in the alloy enhanced hydrogen permeability while maintaining good selectivity. The study shed light on the relationship between alloy composition and membrane performance, suggesting recommendations for enhancing membrane design. Wang et al. created Pd-Zr alloy membranes by a combination of electroless plating and heat treatment and tested their hydrogen separation capabilities. The membranes had good hydrogen permeability and selectivity, which was due to the creation of a dense, defect-free alloy layer with optimal composition and microstructure. The study emphasized the necessity of production procedures and alloy composition optimization in achieving excellent membrane performance. Liu et al. studied the hydrogen separation capability of Pd-Zr alloys formed through mechanical alloying. They discovered that mechanical alloying produced fine-grained microstructures, which resulted in higher hydrogen permeability than conventionally manufactured membranes. The work showed that mechanical alloying is a promising production process for Pd-Zr membranes.

2.2.2.1.7 Palladium-Vanadium (Pd-V) Membranes

Pd-V membranes, a type of metallic membrane, have emerged as a promising candidate for hydrogen separation. They are made up of a thin layer of palladium (Pd) alloyed with vanadium (V) on a porous substrate. Hydrogen separation in Pd-V membranes is accomplished through a mixture of solution-diffusion and surface reaction processes. Hydrogen molecules dissolve in the palladium layer and diffuse across the membrane, whilst other gasses are mostly inhibited. Vanadium atoms in the alloy contribute to higher hydrogen permeance and surface reaction rates.

Lee et al. studied the hydrogen permeation and separation capabilities of Pd-V membranes. The membranes were created by electroless plating Pd-V alloys onto porous stainless steel substrates. The membranes' hydrogen permeance and selectivity rose as the alloy's V concentration increased. At 300°C, the Pd-50%V membrane had the highest hydrogen permeance, measuring 1.1×10^{-7} mol/m^2 s Pa. The membrane's hydrogen selectivity was likewise strong, reaching 100 for a feed combination containing 50% hydrogen. Lee et al. designed high-flux Pd-V alloy membranes for hydrogen separation. The membranes were created by combining electroless plating and thermal annealing. The membranes' hydrogen permeance increased as the alloy's V concentration increased, whereas hydrogen selectivity dropped. At 300°C, the Pd-70%V membrane had the highest hydrogen permeance, measuring 3.2×10^{-7} mol/m^2 s Pa. The membrane's hydrogen selectivity was 50 for a feed combination that contained 50% hydrogen. Kag et al. investigated the effect of vanadium content on the hydrogen separation capabilities of Pd-V membrane. The membranes were created by electroless plating of Pd-V alloys onto porous stainless steel substrates. The membranes' hydrogen permeance and selectivity rose by up to 50% as the V content in the alloy increased. However, increasing the V concentration resulted in a decrease in hydrogen permeation and selectivity. The ideal V content for hydrogen separation was determined to be 50%. Park et al. studied the hydrogen permeation and separation capabilities of Pd-V-Ni membranes. The membranes were created by electroless plating of Pd-V-Ni alloys onto porous stainless steel substrates. The membranes hydrogen permeance and selectivity rose as the alloy's V and Ni concentrations increased. At 300°C, the Pd-30%V-10%Ni membrane had the highest hydrogen permeance, measuring 5.6×10^{-7} mol/m^2 s Pa. The membrane's hydrogen selectivity was 120 in a feed mixture containing 50% hydrogen. Kim and colleagues looked at how stable were Pd-V membranes used in hydrogen separation over time. The membranes were produced by porous stainless steel substrates that were electroless plated with Pd-V alloys. The membranes were subjected to a hydrogen-rich gas combination at 300°C for 1000 hours. Throughout the testing period, the membranes' hydrogen permeance and selectivity remained steady. The membranes demonstrated exceptional stability and durability for hydrogen separation.

2.2.2.1.8 Palladium-Rare Earth (Pd-RE) Membranes

Pd-RE membranes have higher hydrogen permeability than pure palladium membranes. Rare earth elements (REs) increase hydrogen solubility via changing the membrane's electrical structure. REs such as yttrium (Y) and cerium (Ce) create solid solutions with palladium, causing lattice aberrations that promote hydrogen diffusion. Pd-RE membranes' selectivity is determined by the RE's affinity for hydrogen and other gases. REs with a higher affinity for hydrogen, such as Y and Ce, decrease the permeability of other gases, leading to increased hydrogen selectivity. Recent research has focused on refining the composition and structure of Pd-RE membranes to improve their performance. The discovery of new REs, as well as the creation of composite membranes with additional functional layers, hold the prospect of further improving hydrogen permeability and selectivity.

Researchers, including Harris and his colleagues, conducted extensive studies on palladium alloys that involved REs like yttrium, cerium, and gadolinium. They also explored the potential use of these alloys as hydrogen permeation membranes. It was anticipated that palladium and rare earth alloys would demonstrate solid solution hardening and lattice expansion as a result of their notable size disparities. Through careful analysis, the ideal concentrations of these elements in palladium alloys were found to be 5.75% Ce, 8% Gd, and 8% Y (at%). The behavior of cerium during the hydrogen permeation process is greatly influenced by temperature and pressure. Yttrium demonstrates superior performance, while cerium improves permeability in comparison to Pd-Ag at temperatures exceeding 350°C. Reducing differential pressure leads to extra permeability losses, just like Pd-Ag. In contrast to Knapton's research on Pd-Ce (6 at%), these findings demonstrate significant decreases in permeability.

Through extensive research, it has been discovered that when yttrium is subjected to hydrogenation, it remains unchanged in terms of its dimensions. This finding sheds light on the mechanical characteristics of this element. Furthermore, it demonstrates comparable thermal expansion characteristics and increased strength when compared to Pd-Ag alloys. However, yttrium experienced considerable corrosion problems after undergoing oxidative activation procedures. Yttrium underwent preferential oxidation on the alloy's surface, leading to the creation of Y_2O_3. The compound exhibits a relatively weak affinity for the membrane's surface, resulting in the formation of a residual layer enriched with Pd-Y. This is caused by the removal of yttrium from the surface. Scientists have discovered the harmful impacts of carbon monoxide (CO) and hydrocarbons on Pd-Y membranes. With a scientific mindset, Sakamoto delved into the realm of binary alloys, specifically focusing on yttrium and gadolinium. Drawing inspiration from Harris' positive permeability discoveries, Sakamoto embarked on a quest to develop groundbreaking ternary alloys. These tests also used membranes without proper support, and in both cases, there was a notable improvement in permeability, similar to the results mentioned in Harris' articles. In their study, Kang et al. utilized computational methods to uncover new alloying elements that have the potential to improve the hydrogen permeability of Pd-based membranes. The membranes were made up of $Pd_{96}M_4$. These simulations show that thulium and europium are the top choices for improving permeability in alloys. Through experimental testing with a sputtered Pd-Tm membrane, the simulation results were validated and produced extremely positive outcomes. The membrane's permeability showed a 50% increase compared to Pd-Ag and reached twice the permeability value of pure palladium. However, the limited availability and high price of thulium, along with the noticeable hydrogen embrittlement observed during the permeation test, make this binary alloy unattractive as a hydrogen separation membrane. On the other hand, there is a growing need for hydrogen as a sustainable energy source, which is expected to drive the increased use of Pd-RE membranes in hydrogen separation and purification systems in the future. With their outstanding hydrogen permeability, selectivity, and stability, these materials prove to be highly competitive for various industrial applications.

2.2.2.1.9 Palladium-Ceramic Composite Membranes

Palladium-Ceramic composite membranes are a potential technique for the separation and purification of hydrogen. These membranes possess the unique capacity to allow hydrogen to pass through, which is attributed to the hydrogen permeability of palladium. Additionally, they exhibit high mechanical strength and stability due to the presence of ceramic elements. As a result, these membranes offer enhanced performance and durability in comparison to conventional palladium membranes. This introduction will examine the basic principles, manufacturing techniques, and possible uses of palladium-ceramic composite membranes for the purpose of separating and purifying hydrogen. Palladium-ceramic composite membranes provide numerous benefits for the separation and purification of hydrogen in various applications. First and foremost, the incorporation of ceramic materials improves the stability and mechanical strength of the membrane, allowing it to function well under high temperatures and pressures. Additionally, composite membranes demonstrate a notable capacity for both effective separation and high permeability of hydrogen from gas mixtures. Moreover, composite membranes can be customized for certain uses by modifying variables such as the thickness of palladium, the composition of the ceramic, and the structure of the pores. Ceramic materials, including SiO_2, Al_2O_3, ZrO_2, and YSZ, are widely utilized as intermetallic layers.

Lee et al. conducted a study on the advancement of palladium-ceramic composite membranes for the purpose of hydrogen separation. The membranes were created by using electroless plating to deposit palladium onto a porous ceramic substrate. An investigation was conducted to analyze the impact of plating time, temperature, and solution composition on the performance of the membrane. The membrane that was optimized demonstrated a hydrogen permeance of 9.4×10^{-6} mol/m^2/s/Pa, along with a hydrogen selectivity of 550.

Datta et al. conducted an assessment of the efficiency of palladium-ceramic composite membranes in separating hydrogen. The membranes were fabricated by the process of chemical vapor deposition (CVD), where palladium was deposited onto a porous alumina substrate. The membranes demonstrated a hydrogen permeability of 1.3×10^{-6} mol/m^2/s/Pa, along with a hydrogen selectivity of 320. The membranes exhibited excellent thermal stability, maintaining their integrity even after prolonged operation at temperatures as high as 500°C. Hashim et al. conducted a study on the application of palladium-ceramic composite membranes for the process of purifying hydrogen. The membranes were produced using an innovative sol-gel technique. The membranes demonstrated a hydrogen permeance of 2.5×10^{-6} mol/m^2/s/Pa, together with a hydrogen selectivity of 450. The membranes demonstrated efficacy in eliminating contaminants, including carbon monoxide, carbon dioxide, and water vapor, from hydrogen streams. Lee et al. conducted a comprehensive assessment on the advancements made in the creation and utilization of palladium-ceramic composite membranes for the purpose of separating and purifying hydrogen. The discussion revolved around the several aspects that influence the performance of membranes, including their structure, palladium loading, and operating circumstances. The membranes have the potential to be used in hydrogen production, purification, and fuel cell systems. Chen et al. performed an extensive investigation on the enduring stability and durability of palladium-ceramic composite membranes throughout hydrogen purification operations, considering the conditions encountered during operation. The study encompassed accelerated aging experiments, structural analysis, and long-term performance assessment. The results showcased the durability and dependability of the composite membranes, highlighting their capacity for ongoing and environmentally friendly hydrogen separation processes.

In their study, Zhang et al. combined empirical research with mathematical modeling to evaluate the effectiveness of palladium-ceramic composite membranes for hydrogen separation. The study aimed to comprehend the transport mechanisms of hydrogen across the membrane and forecast the membrane's performance under various operational scenarios. The results yielded useful insights into the processes of mass transfer and membrane behavior, hence assisting in the design and optimization of composite membranes for practical applications. Wang et al. examined the production techniques and characterization of palladium-ceramic composite membranes designed for high-temperature hydrogen separation. The study emphasized the impact of factors such as the shape of the ceramic support, the amount of palladium used, and the thickness of the membrane on the capacity of hydrogen to pass through and the ability to selectively separate hydrogen from other gases. The results showed that by optimizing these parameters, the performance of the membrane could be greatly improved. This opens up the possibility of efficiently separating hydrogen at high temperatures. Li et al. conducted a study that specifically examined the production of palladium-ceramic composite membranes utilizing different methods, including electroless plating and physical vapor deposition. The results showed that the hydrogen permeability and selectivity were improved when compared to membranes made of pure palladium. The composite membranes demonstrated enhanced mechanical robustness and durability, rendering them well-suited for utilization in industrial settings.

2.2.2.1.10 Palladium-Carbon Nanotube (Pd-CNT) Membranes

Palladium-Carbon nanotube (Pd-CNT) membranes have attracted considerable interest in recent times due to their potential uses in hydrogen separation and purification. Carbon nanotubes possess a large surface area, impressive mechanical qualities, and chemical stability, making them ideal for supporting palladium nanoparticles. This support enhances the permeability and selectivity of hydrogen. These membranes provide notable benefits such as exceptional durability, tolerance to extreme temperatures, and immunity to contamination, rendering them well-suited for a range of industrial uses, including purifying hydrogen for fuel cells, synthesizing ammonia, and carrying out petrochemical processes.

Palladium demonstrates a strong attraction to hydrogen, since hydrogen atoms easily adhere to the surface of palladium. Carbon nanotubes act as a substrate, improving the distribution and durability of palladium nanoparticles, making it easier for hydrogen to be absorbed. After being adsorbed, hydrogen atoms move through the palladium lattice by a process called solution-diffusion. This movement is driven by the difference in hydrogen partial pressure across the membrane. Hydrogen atoms diffuse across the palladium lattice and undergo dissociation and recombination cycles, resulting in permeation of the membrane. Carbon nanotubes serve to give structural reinforcement and prevent the clustering of palladium nanoparticles, therefore preserving the integrity of the membrane throughout the process of hydrogen penetration. The selectivity of Pd-CNT membranes is due to the preferred adsorption and diffusion of hydrogen compared to other gases, such as nitrogen or methane. The selectivity is determined by various parameters, including the physical characteristics of the membrane, such as pore size, surface shape, and the degree of contact between hydrogen and the membrane material. Palladium-carbon nanotube (Pd-CNT) membranes demonstrate resilience against contamination from substances like sulfur compounds or carbon monoxide, which have the potential to render ordinary palladium membranes inactive. Carbon nanotubes serve to alleviate the consequences of poisoning by acting as a safeguarding shield and facilitating the restoration of active palladium sites.

Li et al. conducted a study on the production of Pd-CNT composite membranes by the utilization of CVD and electroless plating processes. The results demonstrated that the addition of carbon nanotubes to the palladium matrix greatly enhanced the hydrogen permeability and selectivity in comparison to membranes made only of pure palladium. The improved performance was ascribed to the carbon nanotubes' large surface area and structural integrity, which allowed the adsorption and diffusion of hydrogen. Zhang et al. examined how the functionalization of carbon nanotubes affects the efficiency of Pd-CNT membranes in separating hydrogen. The surface chemistry of carbon nanotubes was modified using functionalization procedures, including acid treatment and surface modification. The results indicated that the incorporation of functionalized carbon nanotubes resulted in improved dispersion of palladium nanoparticles and higher stability of the membrane. This, in turn, led to increased hydrogen permeability and selectivity. Wang et al. conducted research on creating and analyzing Pd-CNT composite membranes that are specifically engineered for separating hydrogen at high temperatures. The study examined innovative techniques for creating new compounds and analyzed how different temperatures affected the efficiency of the membrane. The results showed that the Pd-CNT membranes demonstrated exceptional thermal stability and retained a high level of hydrogen permeability even at high temperatures. This indicates that they have great potential for application in industrial hydrogen purification processes. Chen et al. suggested a method of modifying the surface to improve the efficiency of hydrogen separation in Pd-CNT composite membranes. The membranes' surface properties were customized using surface modification techniques, including plasma treatment and chemical functionalization. The study showcased that Pd-CNT membranes with surface modifications show enhanced hydrogen permeability, selectivity, and resistance to poisoning. This makes them highly promising for practical applications in hydrogen purification. For their experiment, Sazali et al. utilized a combination of P84 co-polyimide and nanocrystalline cellulose (NCC) as a precursor to fabricate nanostructured membranes. This approach was selected due to the unique rod-shaped nanostructure and the low decomposition temperature of NCC. We examined different NCC loadings and a range of carbonization temperatures. For optimal results, the membranes underwent carbonization at a temperature of 80°C, along with the incorporation of 7% by mass of NCC. These membranes demonstrated an impressive selectivity of 435 for H_2/N_2. Teixeira and colleagues produced carbon molecular sieve membranes (CMSMs) by combining phenolic resin with boehmite nanoparticles. In particular for larger gas molecules, it was demonstrated that raising the carbon/Al_2O_3 ratio promotes the formation of CMSMs with a more porous structure and higher permeability. The best membranes (M1) for H_2 separation showed a selectivity for H_2/CO_2 that was somewhat near to the Robeson upper limit and an obviously higher selectivity for H_2/N_2. Few researchers have employed the method of doping membranes with metal nanoparticles to enhance the hydrogen separation efficiency of CMSMs. Platinum and palladium are commonly utilized as doping metals because of their strong attraction to hydrogen. Typically, the process of doping with metals involves spreading metal compounds into the membrane precursor prior to carbonization. Nevertheless, alternative techniques for doping subsequent to carbonization have also been devised. As a first step toward CMSMs, Yoda and colleagues produced metal-doped polyimide (PI) sheets by supercritical impregnation using carbon dioxide. As doping agents, platinum and palladium were tested. The Pd-doped CMSMs showed better separation properties and a more homogeneous distribution of metal nanoparticles after carbonization than the Pt-doped membranes. Pd-doped membranes had a much lower permeability to N_2 than did undoped membranes. Permeability to H_2 decreased, however, less noticeably. The Pd-doped CMSM hence showed a far higher H_2/N_2 selectivity of 5640 than an undoped CMSM's selectivity of 330.

2.3 CERAMIC MEMBRANES FOR HYDROGEN SEPARATION

Ceramic membranes are considered highly promising for hydrogen separation because of their exceptional chemical stability, tolerance to high temperatures, and good mechanical qualities. Ceramic membranes possess distinct benefits over traditional polymer or metallic membranes, rendering them highly suitable for challenging hydrogen separation tasks in several industries. Ceramic membranes are employed in hydrogen generation processes, specifically in steam methane reforming (SMR), to selectively isolate hydrogen from the reformed gas stream, hence improving the purity of the hydrogen product. Their ability to withstand high temperatures enables efficient retrieval of hydrogen from high-temperature reforming operations. Ceramic membranes are essential for purifying hydrogen streams obtained from diverse sources such as natural gas, biomass, and coal gasification. Ceramic membranes aid in the production of high-purity hydrogen that is suited for fuel cell applications and chemical synthesis by selectively eliminating contaminants, including carbon dioxide, water vapor, and sulfur compounds. Ceramic membranes provide an energy-efficient method for recovering and recycling hydrogen in industrial processes that involve off-gas streams rich in hydrogen. Ceramic membranes allow for the selective permeation of hydrogen while preventing the passage of other gases. This enables the recovery of hydrogen, which can then be reused. This procedure promotes sustainability and conserves resources. Ceramic membranes are a highly promising technique for separating hydrogen. They have excellent chemical stability, can withstand high temperatures, and have customized pore geometries that allow for efficient and selective penetration of hydrogen. Due to continuous progress in ceramic materials and membrane fabrication techniques, ceramic membranes are positioned to have a substantial impact in meeting the increasing need for clean and sustainable hydrogen energy solutions in various industrial sectors.

Ceramic membranes, mostly composed of silica, are often fabricated using sol-gel or CVD techniques. Additionally, they are proposing membranes for hydrogen purification due to their very simple manufacturing process, cost-effectiveness, and robustness. The problem of pore opening and shrinkage is a significant concern in microporous membranes. In this study, silicon carbide membranes were examined as a potential solution due to their hardness, which is determined by the material's density and the distribution of pore volume. The hydrogen flow rates achieved in silica-based membranes are typically greater than those achieved in metallic, carbon, or zeolite membranes built on tubular alumina supports. This is mostly due to the thinness of the active separation layer, which can be as small as 20–30 nm. The sol-gel technique is a versatile process that enables the creation of continuous layers of different metallic oxides, such as titania or zirconia. These oxides are known for their enhanced stability when exposed to water vapor. By utilizing liquid organometallic alkoxides at room temperature and pressure, the thickness deposited on the commercial support may be modified. This allows for the manipulation of the morphology, tortuosity, and porosity by introducing functionalization agents like organosilanes. CMSMs, which have been extensively researched for gas separation, are commonly synthesized through the pyrolysis of PI membranes. Their primary limitation is their fragility. High-temperature stability is essential for hydrogen generation applications. Recently, the carbon material was enhanced by including mesoporous silica fillers before the pyrolysis process. This modification resulted in increased fluxes without affecting the selectivity. Protonic electronic-conducting membranes, such as Nd_5LnWO_{12}, belong to a new category of ceramic membranes. These membranes allow for the transport of protons across the membrane without the need for any external force, thanks to a chemical potential gradient. These materials have been documented to enhance the stability in settings with high concentrations of CO_2, such as those seen in integrated gasification combined cycle (IGCC) operations.

Seung Hyun Moon and his team conducted a study on the efficiency of asymmetric ceramic membranes in separating hydrogen. The researchers created tubular asymmetric membranes by employing the sol-gel technique with silica and yttria-stabilized zirconia (YSZ) materials. Subsequently, the membranes underwent high-temperature hydrogen permeation testing. The researchers noted that the YSZ-based ceramic membranes demonstrated better hydrogen penetration characteristics in comparison to silica-based membranes, primarily because of their higher ionic conductivity. The asymmetrical membrane construction enabled the selective transport of hydrogen while efficiently preventing the passage of other gases. Furthermore, the membranes exhibited exceptional mechanical durability and resilience to heat, rendering them appropriate for demanding hydrogen separation tasks at elevated temperatures, such as steam reforming and syngas generation. Xinliang Li and his colleagues conducted a study on the efficiency of mesoporous silica ceramic membranes in purifying hydrogen from syngas. The researchers created silica membranes with precise mesopore architectures by employing a sol-gel technique in conjunction with templating chemicals. The membranes were assessed for their ability to separate hydrogen from simulated syngas mixtures containing carbon dioxide and methane. They discovered that the mesoporous silica ceramic membranes displayed a high level of hydrogen permeance and selectivity as a result of their specifically designed pore structures. The hierarchical pore structure enabled efficient hydrogen diffusion while successfully excluding carbon dioxide and methane molecules. The membranes exhibited favorable efficacy and durability in the purification of hydrogen from syngas streams, presenting a viable resolution for the production of environmentally friendly hydrogen in syngas-dependent operations. Toshinori Tsuru and his colleagues examined how the thickness of a membrane and the conditions under which it operates affect the ability of hydrogen to pass through. They used tubular membranes made by the sol-gel process, which were asymmetrical in nature. The membranes underwent testing at different temperature and pressure levels. Tsuru and his team noticed that the hydrogen permeability of the ceramic membranes made from silica increased as the temperature and thickness of the membrane increased. This was because the movement of hydrogen molecules over the membrane's surface was improved. Nevertheless, an excessive amount of thickness resulted in a decrease in selectivity and an increase in resistance to the flow of gas. The study determined the ideal thickness of the membrane and established the optimal working parameters to obtain a high rate of hydrogen permeation, while also ensuring selectivity and mechanical strength. Their discoveries offered a vital understanding for the development and enhancement of silica-based ceramic membranes for effective hydrogen separation in industrial settings.

Li et al. conducted a study on the efficiency of hydrogen separation using ceramic membranes made from Al_2O_3–SiO_2–ZrO_2 precursors. The sol-gel process is utilized to produce synthesized porous ceramic membranes using the aforementioned precursors. The membrane that was optimized demonstrated a hydrogen permeance of 7.1×10^{-7} mol/m^2/s/Pa and a hydrogen/nitrogen selectivity of 307 at a temperature of 800°C. By optimizing the composition of the membrane and adjusting the sintering temperature, a membrane with a well-defined microstructure and significant porosity was achieved. This enhanced the ability of hydrogen molecules to pass through the membrane. Adding ZrO_2 to the membrane improved the membrane's ability to withstand high temperatures and prevented the particles from fusing together. The membrane demonstrated promise for utilization in industrial hydrogen separation applications.

Chen et al. created a hydrogen-selective ceramic membrane for hydrogen purification using a perovskite material called $La^{0.6}$ $Sr^{0.4}FeO^{3-}$ (LSF), which demonstrated excellent performance. The artificially created membrane demonstrated a significant hydrogen permeability of 1.4×10^{-6} mol/m^2/s/Pa and a hydrogen/nitrogen selectivity of 280 at a temperature of 850°C. The LSF perovskite material exhibited a high level of oxygen ion conductivity and mixed ionic-electronic conductivity, allowing for the selective penetration of hydrogen molecules. The membrane's exceptional performance and long-lasting nature make it a promising contender for hydrogen purification applications in industries such as petrochemicals and fuel cells. Deng et al. fabricated hydrogen-selective ceramic membranes by employing a sol-gel technique utilizing a blend of NiO and YSZ powders. The addition of NiO to the YSZ matrix resulted in the formation of a composite material that exhibits both ionic and electronic conductivity, which enhances the ability of hydrogen molecules to pass through. The constructed optimized membrane has enhanced hydrogen permeance, measuring 1.2×10^{-6} mol/m^2/s/Pa at a temperature of 800°C. The improved membrane exhibited a substantial hydrogen flux and demonstrated efficient hydrogen separation from other gases.

2.3.1 Advantages of Ceramic Membranes for Hydrogen Separation

1. Can function optimally at elevated temperatures, often ranging from 500°C to 1000°C, without experiencing substantial performance degradation.
2. Possessing remarkable chemical stability, these materials may endure exposure to caustic gasses and hard working conditions without undergoing any degradation.

3. Can be manufactured with customized pore structures to achieve specific permeability for hydrogen molecules.
4. Enhanced strength and hardness for membranes made from ceramics.
5. Ceramic membranes are resistant to hydrogen embrittlement, unlike metallic membranes.
6. Capable of functioning effectively in various challenging settings, such as extreme temperatures and corrosive surroundings.
7. Provides the opportunity to customize pore shapes, such as microfiltration, ultrafiltration, and nanofiltration, according to specific requirements.

2.3.2 Disadvantages of Ceramic Membranes for Hydrogen Separation

1. Ceramic materials possess an intrinsic brittleness, rendering them vulnerable to cracking or fracturing when subjected to mechanical stress or heat shock.
2. They generally have higher manufacturing costs compared to polymeric or metallic membranes.
3. Polymeric membranes have greater flexibility than other types of membranes. They can conform to uneven surfaces and tolerate bending without sustaining any harm.
4. Prone to fouling, a process in which pollutants build up on the membrane surface or inside the pores, leading to a gradual decrease in membrane function.
5. Ceramic membranes provide customized pore architectures to achieve selective permeability. However, achieving precise control over pore-size distribution might be difficult.
6. The production of ceramic membranes includes a series of stages, including material synthesis, shape, sintering, and surface modification.
7. Expanding the production of ceramic membranes for extensive industrial use can pose difficulties because of constraints in manufacturing technology and equipment.

2.4 POLYMERIC MEMBRANES FOR HYDROGEN SEPARATION

Polymeric membranes have attracted considerable interest for hydrogen separation owing to their cost-effectiveness, scalability, and adaptability. The membranes consist of customized organic polymers that possess certain features, enabling them to selectively allow the passage of hydrogen while preventing the passage of other gases. The process of hydrogen separation over polymeric membranes involves numerous fundamental mechanisms, including diffusion, solubility, and selectivity. Hydrogen molecules pass through the polymeric membrane by diffusing through the polymer matrix. Hydrogen molecules dissolve into the polymer material and travel through the polymer chains from the high-pressure side (feed stream) to the low-pressure side (permeate stream) in a process known as diffusion. Factors such as the polymer structure, molecular weight, and temperature have an influence on the rate of diffusion. Polymeric membranes have variable levels of gas solubility, including hydrogen. Solubility pertains to the capacity of gas molecules to dissolve within the polymer matrix. The concentration gradient across the membrane in hydrogen separation is determined by the solubility of hydrogen in the polymer material, which in turn affects the driving force for hydrogen penetration. Increased solubility of hydrogen in the polymer results in higher rates of hydrogen permeation. Selectivity refers to the polymeric membrane's capacity to selectively allow hydrogen to pass through while blocking other gases that are present in the feed stream. The process of selective permeation is accomplished by using the disparities in the diffusion and solubility characteristics of hydrogen in comparison to other gas molecules, such as carbon dioxide, methane, and nitrogen. Through the optimization of the polymer composition, shape, and processing conditions, it is possible to build polymeric membranes that demonstrate a high level of selectivity for hydrogen separation.

Polymeric membranes are currently the dominant technology for gas separations, and they are undergoing advanced development due to several key aspects. Firstly, due to their ability to be processed and their affordability, they are often compared to inorganic materials. Furthermore, organic membranes that are in the form of hollow fibers or flat sheets can be readily transformed into hollow fiber or spiral-wound modules, allowing for easy scalability. Hollow-fiber membranes are commonly used in industrial settings because they offer a large surface area relative to their volume. Currently, commercial membrane technology units often utilize either hollow fibers made of polysulfone or spiral-wound membranes made of cellulose acetate (CA). To effectively compete with the established conventional separation processes, it is necessary to develop polymer membranes with ultra-high permeance and high selectivity. These membranes should be tailored to separate gases based on their size, shape, and/or chemical properties. The third concern is the stability of the membrane. An optimal membrane material should possess chemical and physical stability under the operating conditions, as well as productivity and selectivity that are determined by the volume and economic aspects of the process. Hence, polymers utilized for gas separation must satisfy various criteria, including (1) excellent mechanical qualities, (2) high resistance to heat and chemical factors, and (3) the ability to withstand plasticization and physical aging, in order to provide optimal durability and longevity of the membrane.

In the 1970s, the first commercially viable use of a membrane gas separation device was achieved for the purpose of purifying hydrogen. Separating hydrogen from extremely supercritical gases, such as CO_2, CH_4, and N_2, is a straightforward task since hydrogen has an exceptionally high diffusion coefficient compared to all other molecules, save helium. Despite hydrogen's unfavorable solubility coefficient, its diffusivity contribution is dominant and results in high total selectivities. A contrasting situation arises when trying to separate CO_2/H_2 mixtures using CO_2-selective membrane materials due to the challenge of achieving high selectivity using H_2-selective polymer membranes. In this scenario, the concentration of H_2 is higher in the retentate while the CO_2 permeates through the membrane. The presence of ethylene oxide groups results in a high level of flexibility, which in turn leads to a weakside-sieving behavior and high diffusion coefficients. These factors directly contribute to a high level of CO_2 permeability. In addition, a cross-linked organic-inorganic reverse-selective membrane was created using a functional oligomer that contains Polyethylene Oxide (PEO) and the epoxy-functional silanes GOTMS (3-glycidyl oxy propyl trimethoxy silane). This membrane exhibited a high CO_2 permeability of 367 Barrer and an appealing CO_2/H_2 selectivity of 8.95 at 3.5 atm and 35°C. In this scenario, whereas diffusion supports the permeation of hydrogen, competitive adsorption promotes the permeation of other molecules. There is a trade-off between permeability and selectivity in polymer membranes, meaning that a membrane with high selectivity tends to have poor permeability, and vice versa. Robeson demonstrated the inverse relationship between permeability and selectivity in 1991 and provided an updated version of this relationship in 2008. The graphs show the upper-bound curve, which is the region where commercially viable materials can be plotted. Glassy polymers are commonly chosen for purifying hydrogen from a gaseous mixture stream because they exhibit superior separation capabilities, specifically in terms of minor variations in molecule size and shape. This makes them more selective compared to rubbery polymers. Polysulfone(PSF) and Polyimides (PI) are glassy polymers that are widely used in industry due to their high gas permeability coefficients and separation factors, as well as their exceptional mechanical qualities, solubility in safe organic solvents, and easy availability in the market. A membrane module at a small size, containing about 150 hollow fibers made of P84 PI, was utilized to separate hydrogen mixtures. The P84 co-polyimide has been identified as one of the most discerning amorphous polymers, possessing exceptional thermal stability and superior mechanical and chemical characteristics. The membrane performance of H_2, N_2, CO, CO_2, and CH_4, as well as the selectivities of H_2/N_2 and H_2/CO mixtures, were examined. The selectivity values for the H_2/N_2 and H_2/CO mixtures were found to be 78 and 60, respectively.

2.4.1 Transportation Mechanism

The primary mode of transportation in polymer membranes for hydrogen separation is solution-diffusion. This process entails the breaking down of hydrogen molecules and their subsequent movement along the polymer chains. The process has three primary stages: hydrogen sorption onto the membrane surface, diffusion of dissolved hydrogen molecules through the polymer, and hydrogen desorption from the membrane surface into the permeate side. The primary factor that propels hydrogen transfer is the disparity in partial pressure of hydrogen on either side of the membrane. Hydrogen molecules first adhere to the surface of the polymer membrane as a result of feeble intermolecular interactions. The sorption process enhances the solubility of hydrogen into the polymer matrix, enabling it to permeate across the membrane via diffusion. After hydrogen molecules are immersed in the polymer matrix, they spread out via the polymer chains, moving from the side with high pressure (feed stream) to the side with low pressure (permeate stream). Factors such as the polymer structure, molecular weight, temperature, and pressure gradient have an impact on the rate of diffusion. Upon reaching the permeate side, hydrogen molecules detach from the polymer surface and enter the gas phase. As a result, the transport process is finished, enabling the collection of purified hydrogen on the permeate side of the membrane.

2.4.2 Performance Evaluation

The performance evaluation of polymer membranes for hydrogen separation entails the assessment of crucial parameters, including permeability, selectivity, long-term stability, and resistance to fouling.

2.4.2.1 Permeability

Permeability quantifies the speed at which hydrogen molecules traverse the membrane under defined operational circumstances. Typically, determination is achieved by permeation experiments, in which the rate of hydrogen flow across the membrane is recorded over some time. Permeability calculations entail the division of the measured flux by the driving power (pressure gradient) and the thickness of the membrane.

2.4.2.2 Selectivity

Selectivity refers to the membrane's capacity to selectively allow hydrogen to pass through it while excluding other gases that may be present in the feed stream. Attaining high-purity hydrogen product streams requires a significant level of selectivity. Selectivity measurements entail the comparison of the permeability of hydrogen with that of other gases, such as carbon dioxide or methane, in the feed stream.

2.4.2.3 Long-Term Stability

Polymer membranes are assessed for their capacity to sustain their performance during prolonged durations of operation. Long-term stability testing entails exposing the membrane to the feed stream for an extended period and observing any alterations in permeability and selectivity over the duration. Consistent performance guarantees dependable functioning and upkeep of the optimum level of separation effectiveness.

2.4.2.4 Resistance to Fouling

Membrane fouling, resulting from the buildup of pollutants or impurities on the surface of the membrane, can gradually impair the function of the membrane. The evaluation of membrane fouling entails the examination of the membrane's ability to resist fouling in practical working settings, as well as the testing of cleaning processes to determine their usefulness in restoring membrane function.

2.5 COMPOSITE MEMBRANES FOR HYDROGEN SEPARATION

Hybrid or composite membrane materials are created by combining different materials to improve the performance and attributes of membranes used for hydrogen separation. These materials provide benefits such as enhanced selectivity, permeability, mechanical robustness, and stability in comparison to conventional membranes made of a single component. Examples of hybrid or composite membrane materials commonly employed for hydrogen separation include:

2.5.1 Mixed Matrix Membranes (MMMs)

Metal matrix composites (MMCs) are formed by incorporating inorganic fillers or nanoparticles into a polymeric matrix, resulting in a composite material that exhibits improved characteristics. Inorganic fillers, such as zeolites, metal-organic frameworks (MOFs), or carbon nanotubes, enhance selectivity and permeability by offering supplementary surface area, pore structure, and adsorption sites for hydrogen molecules. MOFs provide a flexible foundation for customizing membrane characteristics to meet specific separation needs. Pinnau et al. did comprehensive research on MMMs for the purpose of hydrogen separation and purification. It was shown that adding inorganic fillers to polymeric matrices can greatly improve the selectivity of MMMs for hydrogen separation. The inorganic fillers enhance the selective passage of hydrogen by offering more adsorption sites and molecular sieving capabilities, thereby preventing the permeation of other gases such as carbon dioxide, methane, and nitrogen. His research demonstrates that by optimizing the dimensions, configuration, and loading of the inorganic fillers, the transport characteristics of MMMs may be improved, leading to increased hydrogen flows and enhanced separation efficiency.

2.5.2 Polymer-Ceramic Composite Membranes

Polymer-ceramic composite membranes integrate the pliability of polymers with the exceptional selectivity and thermal stability of ceramics. These membranes generally have a polymeric matrix that is infused with ceramic nanoparticles or coatings. The inclusion of the ceramic component improves the membrane's mechanical strength, chemical resistance, and thermal stability, making it well-suited for applications involving the separation of hydrogen at high temperatures.

Ramakrishna et al. conducted a study on the production and analysis of polymer-ceramic composite membranes (PCCMs) used for separating hydrogen. The researchers created PCCMs by incorporating ceramic nanoparticles, such as zeolites or silica, into polymer matrices, such as polyetherimide (PEI) or polysulfone (PSf), through the utilization of a phase inversion technique. The hydrogen permeation properties of the membranes were assessed under different operating situations. The researchers noted that the addition of ceramic nanoparticles to polymer matrices greatly enhanced the membranes' ability to selectively separate and allow the passage of hydrogen. The ceramic nanoparticles increased the surface area, pore structure, and adsorption sites for hydrogen molecules, which improved the selective permeation of hydrogen while preventing the passage of other gases. The PCCMs demonstrated a notable preference for hydrogen while minimizing the transfer of other gases, which positions them as highly favorable options for hydrogen purification purposes.

Sohrab Zendehboudi and his colleagues conducted a study on the efficiency of polymer-ceramic composite membranes in separating hydrogen during membrane-based hydrogen production procedures. The researchers created PCCMs by blending ceramic nanoparticles, such as alumina or titania, with polymer matrices, such as polyvinylidene fluoride (PVDF) or polyvinylamine (PVA), using a solution casting technique. The membranes were assessed for their hydrogen permeability characteristics and structural integrity in conditions of elevated temperature and pressure. The researchers noted that the PCCMs had improved hydrogen permeability and selectivity in comparison to membranes made solely of polymers. The addition of ceramic nanoparticles enhanced the mechanical strength and thermal stability of the membranes by reinforcing the polymer matrix, especially at high temperatures. The PCCMs exhibited exceptional efficiency in the separation of hydrogen from syngas mixtures, displaying both high purity and productivity. This makes them well-suited for utilization in integrated membrane reactor systems for hydrogen generation processes.

Lei Wang and colleagues conducted a study on the progress of polymer-ceramic composite membranes in separating hydrogen from natural gas streams. The researchers created PCCMs by distributing ceramic nanoparticles, such as zeolites or CMSs, into polymer matrices, such as PI or polyamide (PA), using a solution mixing technique. The membranes were assessed for their hydrogen permeability, resilience to chemicals, and durability over extended periods in industrial gas separation applications. The researchers noted that the PCCMs had exceptional hydrogen selectivity and permeability, surpassing the capabilities of traditional polymeric membranes. The incorporation of ceramic nanoparticles improved the membrane's ability to withstand chemical degradation and fouling, hence extending its operational lifespan in challenging settings. The PCCMs demonstrated potential in efficiently and reliably isolating hydrogen from natural gas streams, hence aiding in the advancement of economically viable and eco-friendly hydrogen generation methods.

2.5.3 Metal-Organic Framework (MOF) Membranes

MOFs are structured substances consisting of metal ions or clusters that are coordinated with organic ligands. MOF membranes possess adjustable pore diameters, substantial surface areas, and accurate molecular sieving properties, rendering them highly promising for hydrogen separation. By incorporating MOFs into polymer matrices, it is possible to produce composite membranes that exhibit improved selectivity for hydrogen without compromising their high permeability.

Michael D. Guiver and his colleagues conducted a study on the advancement of membranes based on MOFs for the purpose of separating hydrogen. The researchers created MOF membranes by employing a layer-by-layer deposition approach, in which MOF thin films were cultivated on porous substrates. The membranes were identified based on their capacity to allow hydrogen to pass through and their ability to maintain their structural integrity. The researchers noted that the MOF membranes displayed a strong preference for hydrogen, which can be attributed to the natural molecular sieving characteristics of MOF materials. The porous nature of MOFs facilitated the effective diffusion of hydrogen while impeding the passage of other gases, such as carbon dioxide and methane. The membranes exhibited favorable hydrogen permeability and stability across different operational circumstances, showcasing their potential for applications in hydrogen purification.

Zhimin, together with his co-authors, conducted an examination of the production and effectiveness of MMMs based on MOFs for the purpose of separating hydrogen. The researchers created MMMs by integrating MOF nanoparticles into polymer matrices, specifically PI or polysulfone (PSf), using a solution casting technique. The membranes were assessed for their hydrogen permeability characteristics and durability over an extended period. The researchers noted that the addition of MOF nanoparticles to the polymer matrices greatly improved the membranes' ability to selectively separate hydrogen and increased their permeability. The MOF nanoparticles enhanced the surface area and adsorption capacity for hydrogen molecules, leading to enhanced separation efficiency. The MMMs demonstrated exceptional mechanical robustness and resistance to fouling, rendering them well-suited for large-scale hydrogen purification operations in industrial settings.

Jeonghun Kim and his colleagues conducted a study to evaluate the efficiency of MOF membranes in separating hydrogen in hydrogen generation systems that utilize membranes. The researchers employed a secondary growth technique to produce MOF membranes, whereby MOF crystals were cultivated on porous substrates. The membranes were assessed based on their hydrogen permeability, stability, and ability to be scaled up. The researchers noted that the MOF membranes had remarkable hydrogen selectivity and permeability, surpassing the capabilities of traditional polymeric membranes. The well-organized microporous structure of MOFs allowed for accurate separation of hydrogen molecules based on size while reducing the unwanted mixing of gases. The membranes exhibited exceptional stability and endurance during continuous operation, showcasing their considerable potential for large-scale hydrogen generation applications.

2.5.4 Graphene-Based Membranes

Graphene is a two-dimensional lattice made up of a single sheet of carbon atoms. It has exceptional mechanical strength, great thermal conductivity, and is impermeable to gases. Graphene-based membranes can undergo functionalization or be coupled with polymers to produce composite materials that exhibit enhanced gas separation performance. Graphene oxide membranes have been studied for their ability to purify hydrogen, thanks to their exceptional selectivity and permeability.

Hui Ying Yang and her colleagues conducted a study on the advancement of graphene oxide (GO) membranes for the purpose of hydrogen separation. The researchers created GO membranes by employing a solution-based filtration technique, in which GO sheets were deposited onto porous supports. The membranes were identified based on their hydrogen permeability properties and structural durability. The researchers noticed that the GO membranes displayed a high level of selectivity for hydrogen, which can be attributed to the molecular sieving function of graphene oxide. The exceptional thinness of GO sheets facilitated the effective conveyance of hydrogen molecules while impeding the passage of bigger gas molecules. Jong Hak Kim and his colleagues conducted a study on the production and effectiveness of graphene-based membranes in the process of separating hydrogen. The researchers created graphene oxide (GO) membranes by using vacuum filtration to separate GO particles from a GO solution, and then reducing them using heat processes. The GO membranes that were created had exceptional hydrogen selectivity and permeability, surpassing the capabilities of traditional polymeric membranes. Graphene's atomic thinness enables fast hydrogen molecule movement while efficiently preventing the passage of other gases. Chuan Zhao and his team conducted a study on the advancement of MMMs for hydrogen separation, using graphene as the main component. They created MMMs by adding GO nanosheets to polymer matrices, such as PVDF or polysulfone (PSf), through a solution casting technique. Integrating GO nanosheets into a polymer matrix greatly enhanced the membranes' selectivity and permeability for hydrogen separation. The increased surface area and adjustable pore structure of

GO augmented the adsorption and diffusion of hydrogen molecules, leading to higher separation efficiency.

2.5.5 Hybrid Inorganic Membranes

Hybrid inorganic membranes are formed by combining various types of inorganic materials, such as zeolites, silica, or alumina. This results in the creation of composite membranes that possess specific and customized features. These membranes have benefits such as superior selectivity, excellent thermal stability, and strong resistance to fouling. The researchers' objective is to achieve efficient hydrogen separation with minimal energy consumption and environmental impact by optimizing the composition and structure of hybrid inorganic membranes. Baoxia Mi and her colleagues conducted a study on the advancement of hybrid inorganic membranes for the purpose of hydrogen separation. The researchers created hybrid membranes by integrating zeolite nanoparticles into a silica matrix by a sol-gel technique. The membranes were analyzed to determine their hydrogen permeability capabilities and their capacity to remain stable when exposed to high temperatures and pressures. The hybrid inorganic membranes show a notable preference for hydrogen owing to the molecular sieving phenomenon caused by zeolite nanoparticles. Zeolites possess a microporous structure that enables the efficient diffusion of hydrogen molecules while effectively obstructing bigger gas molecules. Yi Hua Ma and his colleagues conducted a study on the production and effectiveness of hybrid inorganic membranes for separating hydrogen from syngas mixtures. The researchers created hybrid membranes by incorporating CMSs into a ceramic matrix using a templating technique. The hydrogen permeation properties and long-term stability of the membranes were assessed. The hybrid inorganic membranes displayed exceptional hydrogen selectivity owing to the presence of microporous CMSs. The integration of CMS into the ceramic matrix augmented the adsorption and diffusion of hydrogen molecules, leading to enhanced separation efficiency. Nidal Hilal and his colleagues conducted a study on the advancement of hybrid inorganic membranes for the purpose of separating hydrogen from natural gas streams. The researchers created hybrid membranes by integrating MOFs into a silica matrix by a sol-gel technique. The membranes were analyzed to determine their hydrogen permeability capabilities and their ability to resist fouling. The hybrid inorganic membranes shown a notable preference for hydrogen owing to the well-organized microporous structure of MOFs. The integration of MOFs into the silica matrix augmented the adsorption and diffusion of hydrogen molecules, leading to enhanced separation efficiency. The membranes exhibited exceptional resistance to fouling and chemical degradation, rendering them well-suited for hydrogen purification in natural gas processing facilities.

2.6 MECHANISMS OF HYDROGEN PERMEATION

Hydrogen permeation refers to the phenomenon in which hydrogen atoms or molecules move through a substance. It is an essential factor to consider when selecting materials for the storage, transportation, and processing of hydrogen.

2.6.1 The Solution-Diffusion Model

A network of microscopic pores that function as a filter to separate bigger molecules from smaller ones is the best way to characterize a membrane. When the pore width falls to 5 Å or below, the pore model of membrane transport becomes meaningless. The range of the thermal motion of the polymer chains that make up the membrane includes the pore diameter. Permeation has transitioned from being a flow driven by pressure through small holes to becoming a process of diffusion that is regulated by the movement of the polymer chains. The computer simulations have verified that the transition occurs within pore diameters (polymer chain spacings) ranging from 5 to 10 Å. Computer simulation techniques are used to calculate the position of each atom in the membrane element of a polymer. This calculation is done at many intervals to accurately depict the typical thermal movement of the polymer chains. Currently, molecular dynamic simulations cannot accurately estimate permeant diffusion coefficients. However, it is important to note that the underlying process is diffusion rather than pore flow. The solution-diffusion model is a commonly used mechanism to explain the process of hydrogen penetration through membranes, specifically in polymeric and ceramic materials. The process consists of two primary stages: solution and diffusion. During the solution stage, hydrogen molecules permeate into the membrane material, often at the membrane's surface. The dissolving process is frequently propelled by disparities in hydrogen concentration between the feed and permeate sides of the membrane. Subsequently, the hydrogen molecules in solution disperse over the entirety of the membrane material. After being dissolved, hydrogen atoms disperse through the membrane material due to concentration gradients. The rate of diffusion is influenced by various parameters, including the concentration gradient, temperature, thickness of the membrane, and the characteristics of the membrane material. Hydrogen atoms undergo diffusion across the membrane, transitioning from the region of higher concentration (feed side) to the region of lower concentration (permeate side). The solution-diffusion model postulates that the rate-controlling process in hydrogen permeation is the diffusion through the membrane material, rather than the surface adsorption or desorption. This paradigm is specifically relevant to polymeric membranes, in which hydrogen molecules dissolve into the polymer matrix and then diffuse through it. Additionally, the model can be modified to explain the process of hydrogen penetration in different materials, such as ceramic membranes, by applying analogous concepts.

Smith et al. performed empirical investigations on polymeric membranes to examine the behavior of hydrogen penetration. The researchers discovered that the permeability of the membranes was affected by various factors, including the thickness of the membrane, the mobility of the polymer chains, and the solubility of hydrogen. Their findings confirmed the accuracy of the solution-diffusion model, suggesting that the movement of hydrogen molecules through the polymer matrix was mainly controlled by their dissolution and diffusion. Chen and Wang employed molecular modeling approaches to investigate the process of hydrogen permeation in ceramic membranes. The researchers used computational simulations to examine the impact of pore size, surface chemistry, and temperature on the rates at which hydrogen diffuses. Their discoveries yielded vital understanding of the mechanisms that govern hydrogen permeation in ceramic materials and supported the assumptions of the solution-diffusion model. Gupta et al. performed a comparative investigation on several membrane materials, encompassing polymers and ceramics, employing the solution-diffusion model. The researchers examined the correlation between the shape of the membrane, its transport capabilities, and its ability to allow hydrogen to pass through. Their findings emphasized the significance of carefully choosing suitable membrane materials and improving operational parameters in order to improve hydrogen permeability and selectivity. The aforementioned research collectively illustrates the suitability and importance of the solution-diffusion model for comprehending hydrogen penetration in various membrane types. Researchers can create membranes with customized features for individual applications by taking into account parameters such as membrane composition, shape, and operating circumstances. Nevertheless, there are still difficulties in precisely forecasting the behavior of hydrogen penetration in intricate membrane systems, especially when faced with non-ideal circumstances like elevated pressure or the existence of contaminants. Future research should prioritize the improvement of the solution-diffusion model and the investigation of new membrane materials and topologies. This will help overcome the hurdles and fully exploit the capabilities of hydrogen-related technologies.

2.6.2 The Knudsen Diffusion Model

The Knudsen diffusion model is a significant mechanism employed to explain the process of gas permeation through membranes, particularly the penetration of hydrogen. The Knudsen model, named after the Danish physicist Martin Knudsen, is applicable in cases where the average distance traveled by gas molecules between collisions is similar to or greater than the size of the pores in the membrane material. The pressure difference between the feed and permeate sides is considered in the analysis of hydrogen molecule migration across membrane pores by the Knudsen diffusion model. Whereas the solution-diffusion model mostly takes into account the dissolution and diffusion of gas molecules within the membrane material, the Knudsen diffusion model gives priority to the passage of gas through empty spaces or pores inside the membrane (Wey et al., 2020).

The variety of pore sizes within the membrane material is taken into account by the Knudsen diffusion model. From nanometers to micrometers are possible sizes for pores. The distribution of pore sizes influences the general permeability characteristics of the membrane. Knudsen diffusion can occur when the size of pores is larger than the typical distance a hydrogen molecule may travel without running into other molecules. Because the high-pressure feed side and the low-pressure permeate side have different pressures, hydrogen seeps through the membrane. The molecules of hydrogen are propelled across the holes of the membrane by the pressure gradient. The Knudsen diffusion model depends heavily on the mean free path of gas molecules. Between collisions with other molecules, a gas molecule travels an average distance known as its mean free path. When the average distance traveled between collisions is similar to or greater than the dimensions of the openings in the membrane, Knudsen diffusion becomes significant. In such instances, gas molecules have the ability to permeate through the membrane by repeatedly colliding with the walls of the pores, rather than colliding with other gas molecules. Moreover, temperature has an impact on the rate of Knudsen diffusion. Elevated temperatures typically result in enhanced molecular mobility and accelerated diffusion rates. Consequently, the membrane's ability to allow hydrogen to pass through may enhance as temperature rises, assuming that other elements like the integrity of the membrane remain same. Surface interactions between the gas molecules and the pore walls, along with the size of the pores, contribute to Knudsen diffusion. The interactions between gas molecules can impact the average distance they can travel without colliding with other molecules, which in turn affects the rate at which they can pass through the membrane.

The Knudsen diffusion model provides useful insights into the permeation of hydrogen through porous membranes, especially when the average distance traveled by hydrogen molecules is similar to or greater than the size of the membrane pores. Zhang and Li conducted empirical investigations on mesoporous silica membranes to examine the behavior of hydrogen penetration. Their findings demonstrated that Knudsen diffusion exerted a substantial influence on the transportation of hydrogen through the membranes, particularly under low pressures and in membranes with higher pore sizes. Through the examination of the connection between the structure of the membrane, the distribution of pore sizes, and the permeability of hydrogen, they have offered vital understanding regarding the suitability of the Knudsen diffusion model in mesoporous materials. Kim and colleagues utilized molecular dynamics simulations to examine the process of hydrogen permeation through carbon nanotube membranes. Through their computer analysis, they discovered that the transport of hydrogen in nanotube membranes with small diameters was mostly governed by Knudsen diffusion. This occurred when the average distance traveled by hydrogen molecules, known as the mean free path, approached or surpassed the diameter of the tube. These findings emphasized the need of taking into account the impact of pore-size effects and surface interactions while studying hydrogen permeation at the nanoscale. Patel and Gupta performed a theoretical examination of Knudsen diffusion in polymer membranes for the purpose of hydrogen separation in various applications. Their mathematical modeling investigations showed that Knudsen diffusion grew more pronounced in membranes with smaller pore sizes and lower gas pressures. Through the comparison of their theoretical predictions with experimental data, they confirmed the suitability of the Knudsen diffusion model for accurately forecasting hydrogen permeation rates in polymer membranes. The aforementioned observations jointly emphasized the significance of the Knudsen diffusion model in comprehending the process of hydrogen penetration via porous membranes with different pore sizes and materials. Researchers can accurately forecast and enhance hydrogen permeation rates for different membrane systems by taking into account aspects such as pore-size distribution, surface interactions, and operating circumstances. Nevertheless, there are still difficulties in precisely measuring the impact of Knudsen diffusion, especially in intricate membrane configurations and when operating circumstances are not optimum. Future studies should prioritize the improvement of theoretical models, experimental approaches, and computer simulations to gain a better understanding of the function of Knudsen diffusion in hydrogen permeation. This would help in designing and applying membranes more effectively.

2.6.3 Surface Exchange Model

The surface exchange model (SEM) is a theoretical framework employed to explain the process of hydrogen diffusion through solid substances, specifically metals and alloys. This model specifically examines the interplay between hydrogen molecules and the material's surface, taking into account processes including adsorption, desorption, and surface reactions. The SEM includes various fundamental elements (Popov et al., 2018).

2.6.3.1 Adsorption

Adsorption in this context denotes the phenomenon when hydrogen molecules from the gaseous phase adhere to the surface of the substance. The process of adsorption is essential in determining the concentration of hydrogen atoms on the surface of the material, which in turn affects the rate at which hydrogen passes through the substance. The process of hydrogen penetration involves several steps within the adsorption process.

2.6.3.1.1 Surface Interaction

Hydrogen molecules can engage with surface atoms or locations on a material by different forces, including van der Waals forces, hydrogen bonding, or chemical bonding. The surface contacts are influenced by various parameters like surface chemistry, structure, temperature, and pressure, which determine their strength and characteristics.

2.6.3.1.2 Adsorption Sites

The material surface may possess many types of adsorption sites where hydrogen molecules can adhere. These sites may encompass unoccupied surface areas, imperfections, raised edges, or particular atomic locations with elevated reactivity. The presence and ease of access to these adsorption sites affect the rate at which hydrogen is adsorbed and its equilibrium behavior on the surface of the material.

2.6.3.1.3 Adsorption Isotherm

The process of hydrogen molecules adhering to the surface of the material can be described by an adsorption isotherm. This isotherm explains the connection between the amount of hydrogen adsorbed (surface coverage) and the pressure of hydrogen in the gas phase. The adsorption isotherm can display many shapes and behaviors depending on the individual adsorption mechanism, such as physisorption or chemisorption.

2.6.3.1.4 Equilibrium Adsorption

At equilibrium, the rate of hydrogen adsorption onto the material surface reaches a point where it is balanced by the rate of hydrogen desorption, leading to a constant surface coverage of hydrogen atoms. The equilibrium surface coverage is contingent upon variables such as temperature, pressure, and the characteristics of the interaction between the surface and the material. Variations in these parameters can modify the equilibrium adsorption behavior.

2.6.3.1.5 Impact on Permeation

Hydrogen molecules' adsorption onto the surface of the material modifies the gradient of hydrogen atom concentration close to the surface, which modifies the rate of hydrogen permeability through the material. Higher surface coverage of hydrogen adsorbed into a material can increase the driving force for hydrogen diffusion, hence increasing penetration rates.

2.6.3.2 Surface Diffusion

Surface diffusion is the flow of adsorbed hydrogen atoms or molecules along the surface of a material as described in the SEM for hydrogen permeation. In order to help hydrogen atoms move toward permeation or desorption sites, surface diffusion is essential. This in turn affects the overall rate of hydrogen permeation through the solid. Steps under the surface diffusion process are followed in order to penetrate hydrogen.

2.6.3.2.1 Adsorption and Surface Coverage

Hydrogen molecules may occupy adsorption spots on the material's surface or interact with surface atoms or flaws as they adsorb. Surface coverage is the process by which the adsorbed hydrogen atoms deposit a coating or film of hydrogen molecules on the surface. Temperature, pressure, and the kind of surface-material interaction may all affect how much surface is covered.

2.6.3.2.2 Surface Mobility

Surface diffusion is the process by which absorbed hydrogen atoms or molecules travel along a material's surface. Heat energy or temperature gradients cause hydrogen atoms to hop or travel between nearby surface locations, a process known as surface diffusion. A surface's roughness, crystal structure, and defect presence are among the factors that affect hydrogen atom movement.

2.6.3.2.3 Surface Migration

Hydrogen atoms can travel over the surface during surface diffusion and reorganize themselves to get to equilibrium configurations or particular locations like desorption or permeation routes. Surface migration can be mediated by vacancy, interstitial, or coordinated motion of hydrogen atoms along surface steps or defects.

2.6.3.2.4 Rate of Surface Diffusion

Temperature, surface coverage, surface structure, and energy barriers related to atomic movement on the surface all affect the rate of surface diffusion. Generally speaking, higher temperatures cause surface mobility to rise and surface diffusion to speed.

2.6.3.2.5 Effect on Permeation

Hydrogen permeation through a material is determined in large part by surface diffusion. Through effective surface diffusion, hydrogen atoms can be transported more readily to permeation sites, so enhancing penetration through the material. On the other hand, restrictions or obstacles to surface diffusion can reduce the rate of hydrogen permeation generally.

2.6.3.3 Desorption

Hydrogen molecules that have been adsorbing can desorb back into the gas phase. Thermal energy can cause desorption on its own or can be triggered by variables like pressure or temperature changes. Steps under the desorption process allow for hydrogen penetration.

2.6.3.3.1 Equilibrium between Adsorption and Desorption

A layer or film of adsorbed hydrogen can form when hydrogen molecules from the gas phase come into touch with the surface of the material and adsorb onto surface sites or interact with surface atoms. Hydrogen atoms fill the surface in steady-state when the rate of adsorption and desorption equalizes at equilibrium.

2.6.3.3.2 Thermal Energy

Thermal energy drives adsorbed hydrogen molecules to desorb from the surface. With their kinetic energy at finite temperatures, adsorbed hydrogen atoms can break free from the binding forces retaining them to the surface and go back to the gas phase. The higher thermal energy accessible to the adsorbed species makes desorption more likely as temperature rises.

2.6.3.3.3 Pressure and Temperature Dependence

The degree of the hydrogen-material interaction as well as temperature and pressure affect the pace of desorption. Because they make more thermal energy accessible to the deposited hydrogen atoms and lessen the driving force for adsorption, higher temperatures and lower hydrogen pressures typically encourage desorption.

2.6.3.3.4 Surface Coverage

The rate of desorption depends on how much surface area adsorbed hydrogen atoms cover. Because adsorbed species are interacting more at increasing surface coverages, desorption may start to compete with adsorption. On the other hand, increased rates of hydrogen permeability through the material may result from more favorable desorption at lower surface coverages.

2.6.3.3.5 Effect on Permeation

Hydrogen penetration through a material is influenced by desorption on the concentration gradient of hydrogen atoms close to its surface. Reduced surface concentrations of hydrogen atoms brought about by higher rates of desorption encourage hydrogen atom diffusion into the material and increase the rate of permeation generally.

2.6.3.4 Surface Reaction

Surface reactions are those chemical reactions that take place between a material's surface and hydrogen molecules. Different mechanisms, such as dissociation, recombination, or reaction with surface pollutants, can be involved in these reactions.

Hydrogen atom concentration at the material surface and the general rate of hydrogen penetration through the material are significantly influenced by surface reactions. Steps under the surface reaction process are followed to penetrate hydrogen.

2.6.3.4.1 Dissociation

Surface interactions allow hydrogen molecules on a substance to separate into atomic hydrogen. Usually, catalytic surfaces are present or the reaction occurs at high temperatures. Adsorbable species that facilitate surface coverage or hydrogen penetration can be produced by dissociated hydrogen atoms combining with defects or surface atoms.

2.6.3.4.2 Recombination

Recombination is the process by which surface atomic hydrogen species recombine to produce molecular hydrogen, as opposed to dissociation. Surface catalytic sites can either help this process happen naturally or not. Adsorption and desorption processes may be competed with recombination reactions, which also lower the concentration of adsorbed hydrogen atoms on surface.

2.6.3.4.3 Surface Contaminants

Surface interactions can also refer to the interaction of hydrogen molecules with surface pollutants or impurities found on the surface of the substance. Other reactive species, oxides, and sulfur compounds may be among these pollutants. The surface chemistry and the way hydrogen adsorbs, desorbs, and permeates through a material can be modified by surface interactions with impurities.

2.6.3.4.4 Reaction Kinetics

Surface reactions are kinetically determined by things like temperature, pressure, surface composition, and the existence of catalytic sites. The particular surface reaction pathway determines whether the kinetic mechanisms—Langmuir-Hinshelwood, Eley-Rideal, or Mars-van Krevelen—reaction rates follow.

2.6.3.4.5 Effect on Permeation

The rate of hydrogen permeability through a material is influenced by surface reactions on the concentration and distribution of hydrogen atoms at the surface. The type and speed of surface reactions can change the surface coverage, surface chemistry, and surface diffusion kinetics of hydrogen atoms, therefore promoting or inhibiting hydrogen permeation.

2.6.3.5 Permeation Flux

Atoms of hydrogen passing from the feed side to the permeate side of a material are referred to as permeation inflow. Disparities in the partial pressure, pressure, or concentration of hydrogen between the two surfaces of the material drive this process. Surface interactions, adsorption, desorption, and surface diffusion that take place within the material all affect the influx of permeation. The steps under the permeation flux process are followed to penetrate hydrogen.

2.6.3.5.1 Concentration Gradient

A concentration gradient of hydrogen atoms between the material's feed side (high concentration or pressure) and its permeate side (low concentration or pressure) causes permeation inflow. Atoms of hydrogen are propelled to diffuse through the substance by this concentration gradient.

2.6.3.5.2 Pressure Gradient

Apart from concentration gradients, variations in the partial or total pressure of hydrogen between the feed and permeate sides may also be the cause of permeation inflow. Atoms of hydrogen can flow from regions of higher pressure to regions of lower pressure when pressure differentials exist throughout the material.

2.6.3.5.3 Surface Exchange Processes

Hydrogen atom concentration and distribution close to the material surface are influenced by surface reactions, adsorption, desorption, and surface diffusion processes. These surface exchange activities affect the total permeation inflow rate and establish the hydrogen atoms' availability for permeation.

2.6.3.5.4 Surface Coverage and Surface Diffusion

Hydrogen atom flux over the material surface is influenced by the rate of surface diffusion and the degree of surface covering by adsorbed hydrogen atoms. Reduced concentration gradient across the material by higher surface coverages may prevent permeation inflow, whereas effective surface diffusion can increase hydrogen atom transport to permeation sites.

2.6.3.5.5 Material Properties

Additionally, affecting permeation influx are the material's characteristics, including porosity, permeability, surface area, and surface chemistry. Owing to more channels for hydrogen diffusion, materials with higher porosity or surface area may show higher rates of permeation input.

2.6.3.5.6 Temperature Dependence

Temperature often determines the influx of permeation; higher temperatures usually result in faster diffusion rates and higher molecular mobility. But the temperature's impact on permeation inflow can differ based on the particular material and surface exchange mechanisms at work.

Li and Zhang looked into hydrogen permeation behavior experimentally on palladium-based membranes. Their findings showed how crucial surface processes like recombination and dissociation are to regulating the flow of hydrogen across the membrane. Through measurement of surface reaction kinetics and correlation with rates of hydrogen permeation, they offered important new information about the function of surface chemistry in hydrogen permeation. Kim and colleagues studied hydrogen penetration via graphene-based membranes using computational modeling methods. Their simulations showed that, with hydrogen atoms diffusing quickly along graphene surfaces, surface diffusion was a major factor in hydrogen transport. Through an analysis of the impact of surface shape, flaws, and temperature on surface diffusion kinetics, they clarified the processes controlling hydrogen permeation in two-dimensional materials. With the SEM, Gupta and Patel performed theoretical investigations of hydrogen permeability in metal alloys. Their research centered on how surface reactions, adsorption, and desorption interact to determine the hydrogen flow across the alloy membranes. Their mathematical models of surface exchange kinetics gave predictive tools for alloy composition and surface treatment optimization to improve hydrogen permeation performance. These results taken together highlight how important the SEM is for clarifying the processes by which hydrogen permeates through different materials. Researchers can develop materials with customized permeation properties for particular purposes and obtain an understanding of the parameters controlling hydrogen flux by taking surface interactions and exchange processes into account. Accurately modeling and forecasting hydrogen permeation in complex material systems is still difficult, though, especially under less-than-ideal circumstances like high temperatures or reactive surroundings. The SEM and experimental methods should be improved in a future study to overcome these obstacles and fully utilize hydrogen-related technology.

2.7 CONCLUSION

The several kinds of hydrogen membranes and the materials used in their manufacture have been thoroughly examined in this chapter. Regarding hydrogen permeability, selectivity, and stability under various operating circumstances, every kind of membrane—from polymeric to metallic and ceramic—offers unique benefits and drawbacks. Because of their size-based exclusion mechanisms, polymeric membranes—like those based on polymers of intrinsic microporosity (PIMs) or polybenzimidazoles (PBIs)—show extremely high selectivity for hydrogen. But their comparatively poor permeability and vulnerability to deterioration under hard conditions call for continuous study to improve their durability and performance. Excellent hydrogen permeability and impurity resistance are features of metallic membranes, particularly alloys of palladium (Pd) and silver (Pd-Ag). Their high-temperature operation qualifies them for use in high-pressure hydrogen separation applications. Still, problems with material stability in harsh environments and manufacturing costs call for more research. Because they are highly selective and thermally stable, ceramic membranes—such as those made of thick mixed-metal oxides or proton-conducting materials like perovskites—offer encouraging substitutes for hydrogen separation. Still under active study are problems with membrane fabrication methods, durability over time, and economy. All things considered, developments in membrane materials and fabrication methods keep propelling hydrogen separation technology forward and enabling its incorporation into a wide range of industrial processes and sustainable energy applications. To fully utilize hydrogen membranes in advancing sustainable energy solutions and reducing environmental effects, future research should concentrate on resolving current issues, improving performance, and increasing production.

REFERENCES

Al-Mufachi, N.A., Rees, N.V., Steinberger-Wilkens, R., 2015. Hydrogen selective membranes: a review of palladium-based dense metal membranes. *Renewable Sustainable Energy Rev* 47, 540–551.

Chen, C., Shen, L., Lin, H., Zhao, D., Li, B., Chen, B., 2024. Hydrogen-bonded organic frameworks for membrane separation. *Chem Soc Rev* 53, 2738–2760.

Jokar, S.M., Farokhnia, A., Tavakolian, M., Pejman, M., Parvasi, P., Javanmardi, J., Zare, F., Goncalves, M.C., Basile, A., 2023. The recent areas of applicability of palladium based membrane technologies for hydrogen production from methane and natural gas: a review. *Int J Hydrogen Energy* 48, 6451–6476.

Ma, J., Huang, Z., Jia, C., Wang, S., Guo, X., Scott, K., 2016. Palladium-based membranes for hydrogen separation: a review. *J Memb Sci* 514, 383–400.

Pati, S., Jat, R.A., Anand, N.S., Derose, D.J., Karn, K.N., Mukerjee, S.K., Parida, S.C., 2017. Pd-Ag-Cu dense metallic membrane for hydrogen isotope purification and recovery at low pressures. *J Memb Sci* 522, 151–158.

Popov, B.N., Lee, J.-W., Djukic, M.B., 2018. Hydrogen permeation and hydrogen-induced cracking. In *Handbook of Environmental Degradation of Materials*; Kutz, M., Ed. Elsevier, pp. 133–162.

Rahimpour, M.R., Samimi, F., Babapoor, A., Tohidian, T., Mohebi, S., 2017. Palladium membranes applications in reaction systems for hydrogen separation and purification: a review. *Chem Eng Process Process Intensif* 121, 24–49.

Wey, M.-Y., Chen, H.-H., Lin, Y.-T., Tseng, H.-H., 2020. Thin carbon hollow fiber membrane with Knudsen diffusion for hydrogen/alkane separation: effects of hollow fiber module design and gas flow mode. *Int J Hydrogen Energy* 45, 7290–7302.

Zhang, K., Way, J.D., 2017. Palladium-copper membranes for hydrogen separation. *Sep Purif Technol* 186, 39–44.

3 Fabrication Techniques for Hydrogen Separation Membranes

3.1 MEMBRANE FABRICATION TECHNIQUES

Membrane longevity, H_2 permeability, and perm selectivity are critically dependent on the film's porosity and adhesion as well as particle size. The various methods of membrane production affect these features differently (Habib et al., 2021). Consequently, to improve the performance of the thin selective layers, careful choice of fabrication methods should be made. Pd-based membranes can be developed using various methods, including electroless plating (ELP), chemical vapor deposition (CVD), physical vapor deposition (PVD), electroplating, microemulsion technique, liquid-impregnation pore plugging, pyrolysis, solvated metal atom deposition, and high-velocity oxyfuel spraying (HVOF) (Kudapa et al., 2024).

3.1.1 Chemical Vapor Deposition (CVD)

Chemical vapor deposition (CVD) can produce solid material that is both very pure and has great performance. In this method, the substrate is subjected to a volatile precursor, which causes the surface of the substrate to undergo a reaction and decomposition while material is also deposited on the substrate co-deposition. Due to the thin layer deposit that has been made on the surface, the chemical reaction can proceed onto the heated surface. Additionally, the chemical reaction is carried out by a product that has been exhausted out of the chamber together with the unreacted precursor gas (Böke et al., 2016a). Because the material deposits solely on the surface that is heated, CVD has the advantage of reducing the amount of material that is wasted during the deposition process. In addition to this, it has a few drawbacks, including the fact that (i) it necessitates exposure to high temperatures, which can cause the material's properties to deteriorate; (ii) it is extremely difficult to keep the temperature consistent; (iii) it generates volatile byproducts that are expelled from the chamber in the form of gas that is highly corrosive, poisonous, and explosive; (iv) expensive capital and operational costs; (v) poor uniformity of coating; (vi) the use of only inorganic material; and (vii) the characteristic of the predecessor (Böke et al., 2016a).

An illustration of the CVD setup that Itoh et al. used is depicted in the schematic design shown in Figure 3.1. The CVD process involves sublimed organometallic material, swept with inert gas and reacted with hydrogen, resulting in thermal

DOI: 10.1201/9781003590682-3

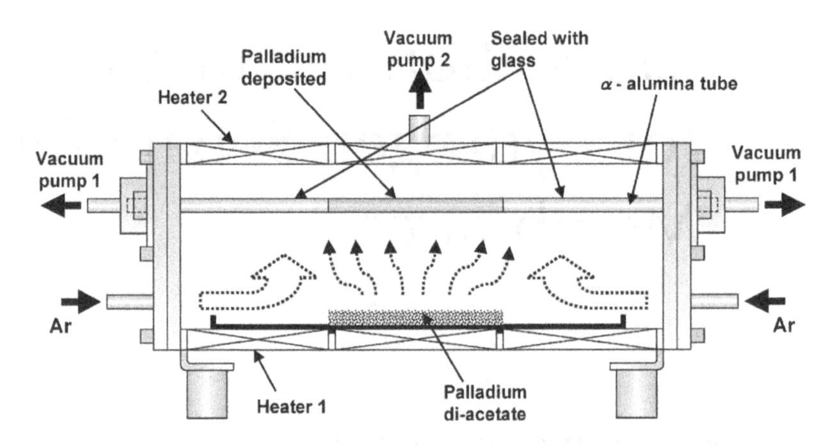

FIGURE 3.1 The equipment used in forced-flow chemical vapor deposition to produce tubular palladium composite membranes (Itoh et al., 2005).

disintegration on support surface. This step is extremely important. This process provides a greater degree of control over the thickness of the manufactured Pd membrane, which is often superior to the control that is available through the ELP method. With the CVD process, it is possible to create a very thin coating that is less than 2 μm in thickness. Ye et al. were the first to report the CVD process as a method for the manufacture of composite Pd membranes. $PdCl_2$ was utilized as a metal source to manufacture the membrane, and a disc of Al_2O_3 was utilized as a support for the layer of Pd material used in the membrane. After some time had passed, Itoh et al. proposed an additional apparatus that utilized the forced-flow CVD process. The system that has been proposed can be used for the creation of Pd composite membranes that possess tubular forms and have selective layers that range from 2 to 4 μm, providing a selectivity of 5000 for H_2/N_2. The rate of external transport or surface hydrogen dissociation becomes the distinguishing element for the films produced by the CVD technique because of the fast passage through the thin Pd layer. To fabricate extremely selective thin layers, the CVD approach necessitates the utilization of palladium precursors that possess both high volatility and a high degree of thermal stability. It is necessary to possess these qualities to achieve a greater yield and a shorter processing time (Böke et al., 2016b).

Even though organometallic Pd precursors with extremely volatile properties (such $Pd(C_3H_5)_2$, $Pd(C_3H_5)(C_5H_5)$, and $Pd(C_5H_5)_2$) can be utilized for the manufacture of palladium selective membranes, the cost of these precursors is significantly higher. Furthermore, the metal-organic chemical vapor deposition (MOCVD) technique necessitates a greater vacuum level (below 0.1 mmHg) as an operating condition. Furthermore, the residual carbon that is produced throughout the process has the potential to contaminate the platinum membranes. The MOCVD method is characterized by some characteristics that have the potential to greatly impact the deposition process. These parameters include the temperature and quality of the substrate, the quantity and quality of the precursor, and the amount of hydrogen present in the environment during the deposition process. Initiated CVD (iCVD), plasma-enhanced CVD (PECVD), metal-organic CVD (MOCVD), aerosol-assisted CVD (AACVD), and atomic layer CVD (ALCVD) are some of the CVD techniques that have been utilized over the years to fabricate thin film membranes. A chemical inhibitor is utilized in the iCVD approach to kickstart the polymerization process. On the other hand, in the PECVD technique, free plasma radicals are utilized to kickstart the plasma deposition process.

Palladium (Pd) membranes that are gas-tight and thin were manufactured by G. Xomeritakis and colleagues using the counter-diffusion CVD procedure. The Pd precursors used in this process were palladium chloride ($PdCl_2$) vapor and hydrogen. The reaction temperature, the concentrations of the reactants, and the quality of the top layer were the most significant criteria that influenced the characteristics of the CVD process. A reduction in the He permeance of the porous substrate of between 100 and 1000 times was observed when platinum was placed in the top layer of $\gamma-Al_2O_3$. The H_2 penetration flux of these membranes was investigated to be between 0.5 and1.0×10^{-6} mol / m^2 / s / Pa between 350°C and 450°C.

Liu et al. (2019) developed a unique method for fabricating hydrogen-selective membranes utilizing CVD techniques. This method involves optimizing deposition parameters such as temperature, pressure, and precursor concentration. The primary objective of their research was to improve the selectivity and permeability of the membranes by optimizing the deposition process. They discovered that by manipulating the growth parameters, such as temperature, pressure, and precursor concentration, membranes could be produced that had superior performance in terms of hydrogen separation. Wang et al. (2020) evaluated the impact that various catalysts have on the manufacturing of hydrogen membranes using CVD. They compared the performance of nickel, palladium, and platinum catalysts based on their findings. The performance of membranes that were manufactured using a variety of catalysts, including nickel, palladium, and platinum, was analyzed and compared in their analysis. The researchers observed that the selection of the catalyst had a substantial impact on the membrane structure as well as the hydrogen permeability properties. A study conducted by Zhang et al. (2018) investigated the possibility of using carbon-based nanomaterials as substrates for the construction of hydrogen membranes using CVD. Nanomaterials, in particular, carbon-based nanotubes and graphene, were investigated for their potential use as substrates in the construction of hydrogen membranes using CVD. As a result of their high surface area and structural features, their studies indicated that nanomaterial substrates have the potential to improve membrane stability and hydrogen permeability. Yamazaki et al. (2017) evaluated the effect that the temperature of deposition and the content of the precursor gas had on the properties of hydrogen-selective membranes that were created via CVD. They found that increasing the temperature at which the deposition took place resulted in denser membranes that had increased hydrogen selectivity values. According to the findings of the analysis, the composition of the precursor gas affected the morphology of the membrane as well as the properties of hydrogen permeation. In their 2016 study, Lee et al. investigated several surface modification strategies to improve the development and characteristics of hydrogen membranes that were deposited via

CVD. The results of their investigations demonstrated that surface functionalization increased membrane quality by increasing the adherence and nucleation of membrane materials. Within the context of optimizing membrane performance, the analysis revealed the importance that surface modification plays (Khatib and Oyama, 2013; Ma et al., 2005).

3.2 PHYSICAL VAPOR DEPOSITION (PVD)

The process of depositing thin films and coatings on the surface is accomplished using a vacuum deposition technique. There are four stages involved in this method: the first is evaporation, the second is transportation, the third is reaction, and the fourth is deposition. An extremely thin coating is formed on the surface of this material after it transitions from the condensate phase to the vapor phase and then back again. PVD has several benefits, including the fact that the deposited material can have better properties than the substrate material, that it is more environmentally friendly than CVD because it does not emit toxic gases, that it can be used with almost all types of inorganic material and some organic materials, and that it is safer than CVD because it does not remove any byproduct gases during the process. On the other hand, it has a few drawbacks, including the following: (i) a high capital cost; (ii) a rate of film deposition that is rather sluggish; and (iii) the requirement of a significant quantity of heat and vacuum for the deposition of solid material on the surface (Böke et al., 2016a).

PVD, or physical vapor deposition, is a vaporization coating technique-based process for producing Pd-based membranes. By means of a high-energy source and the bombardment of a solid metal precursor, this method enables atomic material transfer. The energy source might be an electron beam or ions in a vacuum, and the precursors are the target. Another name for the target is the objective. This is not the case with the CVD method, which usually consists of no chemical breakdown processes. The PVD method uses pure metallic precursors, which are elemental in nature; the CVD method uses vaporized chemical compounds as precursors. Mercury membranes are made by magnetron sputtering, a technique that involves colliding and atomizing the target with argon (Ar) ions. The basis of this method is the PVD technology. These Ar ions get their excitation from the plasma activity. Following its transportation to the substrate, the atomized target is deposited on its surface by the next step. More uniform layers than one micron in thickness can be produced by the very attractive magnetron sputtering technique of membrane production. Furthermore, PVD-magnetron sputtering allows for the efficient control of the coating's microstructure and composition. It is quite challenging to maintain this control composition at an Ag content as high as 23% for the Pd–Ag membrane utilizing other coating techniques as electroless deposition. Besides, this composition is highly maintainable. Since PVD produces no liquid waste from the chemical baths, it is a more ecologically benign technique than electroless plating. PVD is therefore a more ecologically benign method (Pal et al., 2020). Furthermore, the PVD technique gives better control over the composition, thickness, and phase of the film. Photovoltaic (PVD) technologies involve a wide range of techniques, magnetron sputtering being only one. Thermal evaporation and pulsed laser evaporation are two further methods that belong in this group. One of the commonest methods of PVD following sputtering is thermal evaporation. This is so that the process is really simple. Thermal evaporation is the process by which coating material is heated and subsequently evaporates inside a vacuum-filled container. After that, a thin coating of these vaporized coating ingredients is applied to the substrate. The simplicity of this method accounts for its much faster (10^{-3} g/cm^2 s) speed when compared to the other PVD techniques. Still, these methods are inefficient and require more costly machinery to produce a high vacuum with a very high power density in order to evaporate the desired materials. Besides, another element that might be regarded as a limiting component is the geometry. This is because the range of practical applications is thereafter limited as the thin membrane can only be created over flat substrates (Pal et al., 2020).

Athayde and colleagues were able to increase the selectivity of typical polymer membranes by use of the sputter deposition method without appreciably reducing their permeability. On polymeric substrates, they created ultra-thin 50 nm Pd-Ag alloy films. Since the thermal stability of the support limits the operation of these polymer/metal composites, Jayaraman et al. improved this idea by using porous ceramic supports to allow higher temperature operation. While creating gas-tight membranes that were appropriate for particular applications required surface roughness of the support, the preparation of metal films was quick and easy to handle. Another approach to the synthesis of thin supported metal membranes has been reported to be spray pyrolysis.

3.3 ELECTROLESS PLATING (ELP)

The ELP approach depends on the uniform application of metal onto the surface of the target through the process of autocatalysis. Due to the absence of any external sources like electricity or electrodes, a thin metal sheet forms on the substrate, resulting in reduced operational costs. ELP can create a uniform film on intricate structures. The membrane manufacturing process experiences a significant number of rejections, primarily caused by the presence of flaws. This leads to a substantial increase in the overall cost of the membrane. This is in addition to all the existing benefits. Hence, it is vital to utilize an alternative approach to address this problem. Due to its autocatalytic nature, the ELP process exhibits similar behavior. Due to this rationale, ELP has been regarded as a potential technology throughout the past 10 years. The reason for this is that it offers a greater degree of uniformity in the films, in comparison to their equivalents, CVD and PVD. Electroplating deposition, also referred to as EPD, is an additional method that can be employed in the manufacturing of palladium-based membranes (Guo et al., 2014). The electroplating deposition (EPD) is a technique performed in a liquid medium. This method involves the application of electric potential to induce the motion of metal ions, resulting in their deposition onto the support surface. When employing this method, the material on which the membrane is placed is used as an electrode (sometimes referred to as a cathode) for the electrochemical system. In addition, the metal coating on the membrane support is created through the deposition of positively charged metallic ions found in the liquid solution. Despite the electrochemical nature of this method, the equipment needed for membrane production is relatively simple. Through control

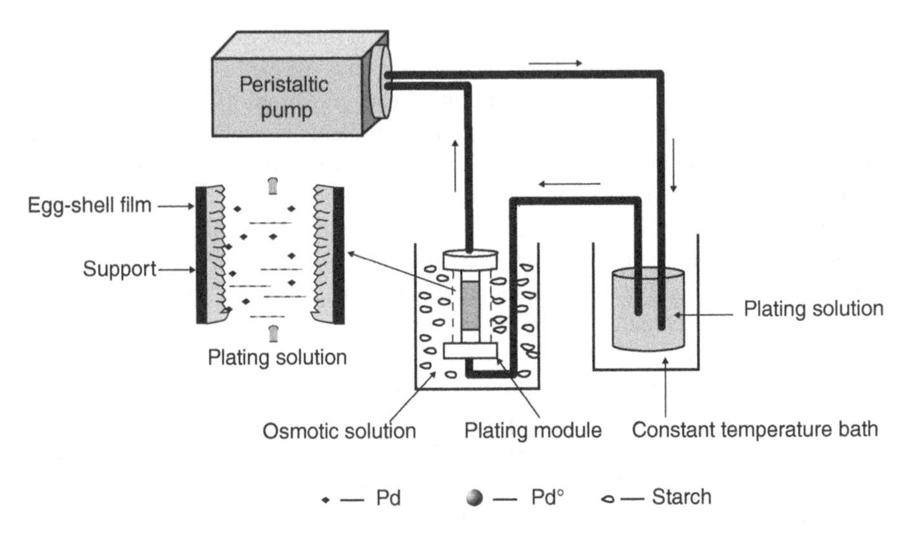

FIGURE 3.2 The process of fabricating palladium membranes utilizing the BELP method (Guo et al., 2014).

of the current density and deposition duration, the EPD technique allows for the modification of the thickness of the selective layer (Figure 3.2). Application of this technique is limited to electrically conductive supports since the EPD procedure is limited to stainless steel or other conductive support materials. Although the EPD process is a preferable choice compared to other methods of membrane production, it can only be carried out under specific conditions (Guo et al., 2014). Although the CVD and PVD techniques can create thin alloy coatings at the nanoscale, they typically require a highly regulated environment. This is a substantial drawback for the manufacturing of products on a big scale. Due to its exceptional repeatability and cost-effectiveness, the electroplating deposition technique is suitable for producing membranes of varying sizes and shapes. However, despite the advantages mentioned before, the ELP technique also has some disadvantages. These factors encompass the challenge of regulating the thickness of the film, the intricacy of the pre-treatment procedure, and the limitation of its use to conduct substrates. It is important to realize that, depending on the kind and condition of the membrane or substrate, it may be feasible to bypass the intricate pre-treatment process in specific situations (Martinez-Diaz et al., 2021).

The utilization of ELP technology for incorporating palladium onto porous substrates has been extensively employed in the production of hydrogen-selective membranes for a considerable period. Because there are no electrodes or external power sources required, this method does not necessitate expensive equipment or result in large operational expenses. This was one of the previously discussed points in the introduction. In addition, ELP can produce uniform films on intricate shapes and non-conductive materials, making it the preferred choice over other technologies. This section provides a comprehensive summary of the method, highlighting the major developments made in recent years using both ceramic and metallic supports.

Palladium (or comparable alloying elements) is applied electrochemically (ELP) to a particular surface by use of an aqueous solution containing the metal precursor. Usually, the precursor is dissolved and stabilized with a ligand to create a complex, then reduced by a carefully regulated autocatalytic chemical process. Ammonium hydroxide and ethylenediaminetetraacetic acid have been commonly used together in many published studies to form a complex with the palladium precursor. But since it produces nitrogen as a special chemical process byproduct, hydrazine is the better reducing agent. This enables it to prevent the deposition of other prohibited substances, such as phosphorous, in the film. Hydrazine is a highly efficient reducing agent that may be utilized in both acidic and alkaline conditions. The reaction circumstances play a crucial role in determining whether it is feasible to convert metal ions with higher valencies into metal ions with lower valencies or to the zero valent state. Below is a detailed explanation of the main chemical processes that occur during the process of palladium deposition:

$$2\ \mathrm{Pd}\left(\mathrm{NH}_3\right)_4^{2+} + 4\mathrm{e}^- \rightarrow 2\mathrm{Pd}^0 + 8\ \mathrm{NH}_3\ \ E^0 = 0.95\ \mathrm{V}$$

$$\mathrm{N}_2\mathrm{H}_4 + 4\ \mathrm{OH}^- \rightarrow \mathrm{N}_2 + 4\mathrm{H}_2\mathrm{O} + 4\mathrm{e}^-\ \ E^0 = 1.12\ \mathrm{V}$$

The combined equation is represented by.

$$2\ \mathrm{Pd}\left(\mathrm{NH}_3\right)_4^{2+} + \mathrm{N}_2\mathrm{H}_4 + 4\ \mathrm{OH}^- \rightarrow 2\mathrm{Pd}^0 + 8\ \mathrm{NH}_3 + \mathrm{N}_2 + 4\mathrm{H}_2\mathrm{O}E^0 = 2.07\ \mathrm{V}$$

Pre-treating the supports by introducing the first nano-sized Pd nuclei before the main plating process is essential to obtain a uniform Pd coating, high adhesion, and the right moment of induction to start chemical reactions. It is vital to obtain the intended outcomes. Historically, this procedure has been carried out by a method known as sensitization-activation therapy. This technique entails multiple immersions in acidic solutions that contain palladium and tin. However, certain studies indicate that the presence of tin residues is responsible for issues related to the integrity of membranes when exposed to elevated operating temperatures. As first observed by Paglieri et al. in the late 1990s and more recently confirmed by other scientists, these residues cause defects and pinholes to develop in the Pd film. A detailed investigation on the relationship between the presence of tin residues and membrane durability has been reported by Wei et al. Other techniques have been suggested in lieu of using tin solutions because of the acknowledged drawbacks associated with traditional sensitization-activation procedures. Activated particles containing Pd nuclei have been used to form intermediate layers; catalyzed anodic alumina surfaces have been used to support Pd electroless plating; ultrasounds have been used to accelerate the deposition of nuclei and break up conglomerates; a palladium acetate solution in chloroform has been incorporated, broken down, and reduced onto the surface; and nano-sized Pd particles have been directly produced by reducing a highly diluted solution with a combination of ammonia and hydrazine. No solution has come out as better up to now, and many experts still think the old approach is the best one (Martinez-Diaz et al., 2021).

Before every electroless plating operation, Uemiya and colleagues submerged the porous support in a hydrazine solution to accelerate the rate of metal assimilation into the material. Several more writers tried to increase the amount of palladium incorporated in deep areas of the surface. Metal particles that obstruct the support's apertures and, by the bridge mechanism, form a totally solid, uninterrupted layer set apart the deep regions. As compared to conventional electroless plating, Zhao et al. and Zhang et al. reported achieving a homogeneous microstructure of the Pd layer with a reduced average thickness by utilizing vacuum on the inner side of the supports. Similar results have been obtained by researchers such as Yeung, Souleimanova, or Li when an aqueous sucrose solution and the plating solution experience an osmotic effect. Pacheco Tanaka and her associates went one step further and made supporting membranes. The membranes were made by sandwiching two ceramic layers of zirconia oxide together using a vacuum-assisted electroless plating process. A Pd seed had already activated one of the layers, which was then placed onto a tubular alumina support. In this way, Pd or alloys based on Pd were included. The selective layer of this membrane, which is called a pore-filled type membrane, is arranged inside a sandwich-like structure. The authors claim that using the membrane below the threshold temperature and preserving total mechanical stability are the two primary advantages. In comparison, membranes that were previously supported and depended on a traditional outer coating occasionally suffered irreversible degradation. Moreover, in case of possible poisoning, the sandwich design gives the selected chemical an additional degree of protection, escalating the damage. On the other hand, some research are now concentrated on changing the plating bath composition in order to enhance the final properties of the palladium layer. Here, it has been demonstrated that conventional electroless plating solutions containing ethylenediaminetetraacetic acid (EDTA) behave consistently at different temperatures. The cleanliness of the palladium layer is impacted, nevertheless, by the carbon residues from the EDTA complex found in the metal particles. The effectiveness of the membrane may be impacted if carbon deposits accumulate and, in particular operating conditions, carbon dioxide is produced. Researchers have therefore also looked into the creation of free-EDTA baths and found that it is possible to maintain the stability of the plating bath while achieving acceptable rates of palladium deposition without the use of these stabilizing chemicals (Alique et al., 2018).

The impact of fluid dynamics on the interaction between the support and the plating bath has been studied by other writers. Chi et al. claim that revolving the support during electroless plating has been demonstrated to increase the plating rate and Pd layer uniformity. This technique resulted in Pd membranes with more even and smoother surfaces, leading to increased stability of the supported system. The authors found that static ELP had a membrane permeance of $3.0 \times 10^{-3} \text{mol} / \text{m} / \text{s} / \text{Pa}$, which is significantly higher than the experimental results. At a thickness of 5 μm, a temperature of 400°C, and a pressure of 4 bar, the hydrogen separation factor was found to be greater than 400.

Although there have been considerable research efforts to improve the quality and cost-effectiveness of Pd membranes, a substantial number of studies focus on minimizing the rejection of membranes caused by flaws or cracks that occur during manufacturing processes. Recent reports have revealed the presence of new options to rectify any defects that have developed on the surface of the Pd. Li et al. utilize the previously established principles of the osmotic effect to prioritize the inclusion of Pd particles in fault locations to repair the Pd layer. Following this approach ensures the removal of every defect, which raises the ideal hydrogen separation factor significantly. This phenomenon occurs without any discernible thickness increase or decrease in the permeation flow. Point plating has been suggested by Zeng et al. as a method of repairing defects in supported Pd-based membranes. Similar ideas underpin this approach. In this instance, flaws are selectively targeted by feeding the metal source and hydrazine solutions from opposing sides of the supporting membrane, hence initiating the chemical process for palladium reduction. We accomplish this while the membrane is still intact. Other researchers have lately reported the isolation of the supply of a Pd source and the solution of a reducing agent to build Pd-based membranes directly on coarse commercial porous stainless steel (PSS) supports after these repair processes. By using the support wall, the electroless pore-plating (ELP-PP) technique keeps the Pd source and hydrazine solutions physically apart. Under such circumstances, hydrazine reacts with the amino-palladium complex close to the pore region after selectively passing through the support's pores. The reduction process starts from the support's internal porosity under ideal circumstances as soon as the inner pore surface is sufficiently stimulated. Comparable to the earlier suggested sealing method, this implies that it is possible to maintain the palladium source and reduce the amount of rejected membranes by using this method. This approach finally results in a reduction in the membrane preparation costs overall. By preventing reactant interaction, Pd inclusion in the material finally completely blocks the pores and ends the process. This approach prevents the rise in palladium incorporation that follows when the pores close, unlike conventional ELP. So, the final film is exceedingly thin and compact (Omidifar et al., 2024).

The scientists discovered that a considerable range of pore sizes present in both conventional and customized PSS supports caused an outer layer to build on them. This was the situation even if Pd was preferentially incorporated into the pores of the support. The hydrazine is unable to permeate the smallest pores due to the full obstruction caused by palladium particles within a relatively brief period. However, the reducing agent can permeate through the largest pores, even if they are partially obstructed until it reaches the outer surface where it hits the palladium bath and forms the exterior layer. Several factors affect the ELP-PP process, including the average pore diameter and porosity of the support, the composition of the reduction and metal plating baths, and the membrane length-to-solution volume ratio. These factors have a significant impact on the ELP-PP process. Smaller holed supports—which are obtained by directly oxidizing commercial supports in the air—are used to make membranes that appear to be roughly 10 μm thick. This number is equal to half of the value obtained when using unaltered supports, which results in a thickness of 20 μm. However, the estimation does not consider the discrepancy between the apparent thickness, determined through gravimetric analysis, and the true value obtained through SEM characterization, which is 2–6 μm more. When put to several simulated and real-world operating conditions in a WGS membrane reactor, the membranes showed remarkable endurance. This novel method has been shown recently to generate supported membranes on ceramic supports with much smaller pores than traditional PSS supports. This emphasizes how much support characteristics affect plating efficacy, with special attention to pore-size distribution and average. Thinner membranes were achieved with this method, despite the presence of palladium in the pores of both the internal and external surfaces.

Using bio-membrane-assisted electroless plating in conjunction with the osmosis method (BELP), Yu et al. effectively created a palladium membrane on the inside surface of an alumina tube. An osmotic mechanism was established for the synthesis of the palladium membrane by the semipermeable egg-shell film. It also served as a shield, keeping the osmotic fluid from contaminating the palladium membrane. Moreover, to improve the mass transfer taking place at the interface between the plating surface and the solution, the plating solution was pumped through the tube side. At 773 K, the 2 μm thick palladium layer of the palladium composite membrane showed a remarkable hydrogen permeation flux of 0.22 mol / m^2 / s and a perm selectivity of 2020 for H_2/N_2. In addition, the membrane demonstrated durability during continuous operation for 240 hours and remained stable after 10 cycles of fast temperature changes between 673°C and 773°C. When employing the osmosis method, it is asserted that the use of an egg-shell layer on the support's surface enables the deposition of a palladium membrane with a larger pore size (Xiong et al., 2022).

3.4 ATOMIC LAYER DEPOSITION (ALD)

The high aspect ratio of thin films deposited on huge surface areas is possible with the ALP technique, which is derived from CVD. By this method, the substrate's surface is exposed to the chemical precursors. The surface group and chemical precursors react self-saturating to produce an ultra-thin film directly. This film is self-limited and is created layer by layer during the entire process. This technique involved two distinct stages: the adsorption and oxidation of metal precursors. This technology has numerous advantages, such as exceptional conformity, consistent performance, the capacity to easily adjust the membrane thickness at the nanoscale level, and a simple and uncomplicated process. Furthermore, this technology is not economically efficient due to several factors, such as the significant quantity of material wastage, the substantial energy consumption, the labor-intensive nature of the process, and the release of nanoparticles. Moreover, this technique is extremely time-consuming because the duration required for chemical reactions is too lengthy. Weber et al. successfully produced a hydrogen-selective membrane using the ALD process in their first prototype. Based on this analysis, the researchers discovered the presence of lead within the pores of the outermost layer of y-Al_2O_3, which coated the inner surface of the ceramic tubular support (Weber et al., 2020).

ALD coatings have been utilized on various membrane substrates, such as porous polymers and inorganic ceramic templated substrates, among others. While most documented research tries to change the surface physicochemical properties of the pores and lower their width, a few have focused on building separative layers with adjustable features. An initial investigation on atomic layer deposition (ALD) on porous membranes was carried out in 1995 by Kim and Gavalas. Applying a thin SiO_2 coating onto a mesoporous Vycor glass tube substrate allowed the researchers to create H_2-selective silica membranes. Subsequently, the reactants $SiCl_4$ and H_2O vapor were deposited in alternating cycles, potentially up to a maximum of 35 cycles. The initial analysis of gas permeance using ALD-modified membranes revealed a decrease of around 20% in H_2 permeance, while their selectivity reached values above 2,000 at a temperature of 600°C. Ducso et al. did the initial work on conformality in porous membranes in 1996. The researchers documented the application of conformal ALD of SnO_2 coatings inside the 65-nm wide pores of porous silicon, which had an aspect ratio (AR) of 1:140. Conformality within porous membranes was explored for the first time. In the same year, Ott et al. utilized ALD to coat the pores of porous ceramic alumina membranes with Al_2O_3 films. The number of ALD cycles was shown to be a relatively linear function of the pore diameters by the liquid-liquid displacement porosimetry method. The decrease took place at a rate of around 0.9 Å per cycle, which aligns with a growth rate of about 0.45 Å per cycle on the surface of the pore (Weber et al., 2020).

Romero et al. conducted ionic transport experiments using nanoporous anodic alumina (AAO) membranes. SiO_2 sheets were created using ALD. The researchers found that the pore radii and membrane porosity were reduced by using ALD to uniformly deposit a thin coating of SiO_2 to the AAO surface and inner pore walls. Furthermore, this activity affected the surface electric fixed charge of the membrane. One may see a clear correlation between the NaCl diffusion coefficient and the effective fixed charge of the membrane. A thin layer of SiO_2 applied to the AAO membranes reduced the positive effective fixed charge by 75%. The decline occurred independent of the membrane pore radii or the increase in counter-ion transport. Using ALD, Xiong et al. characterized AAO membranes both before and after covering them with either TiO_2 or Al_2O_3 to the whole surface, including the interior pore walls. The researchers reported that the AAO membranes demonstrated remarkable thermal stability. Cameron et al. discovered that utilizing ALD on SiO_2 and TiO_2 films led to a decrease in the diameters of the pores in commercially available mesoporous alumina tubular membranes. This study was the inaugural endeavor to achieve this objective. The SiO_2 and TiO_2 ALD layers gradually shrank the membrane pores from their starting size of 50 to molecular dimensions, according to the N_2 permeance measurements. Moreover, it has been found that the movement of gas via tiny openings is controlled not only by the size of the openings but also by the interactions between the transported substances and the material on the surface of the openings. Li and colleagues next employed 50 nm average pore-size ceramic ultrafiltration (UF) membranes as substrates. They proceeded to apply a coating of Al_2O_3 on these membranes using ALD. By adjusting the number of ALD cycles, the pore size of these ceramic ultrafiltration (UF) membranes may be easily modified. This, in turn, enables the adjustment of water permeability. It was discovered that augmenting the number of ALD cycles led to a reduction in water permeability while concurrently enhancing the retention rate. Shang and colleagues studied the use of ALD of titanium dioxide to modify ceramic nanofiltration (NF) membranes. Their low molecular weight cut-off (MWCO) ranges from 260 to 380, and they considerably reduced the size of active holes. Furthermore, they exhibited a notable ability to sustain a comparatively elevated water permeability ranging from 11 to16 $L/m^2/h/bar$. Chen et al. recently conducted a study where they employed ALD of TiO_2 to enhance the tightness of ceramic ultrafiltration (UF) membranes, to produce nanofiltration (NF) membranes. The experiment maintained a constant water permeability of 32 $L/m^2/h/bar$. Moreover, the formed membranes showed a much improved capacity to reject negatively charged dyes when used to remove colors from water. Applying an ALD SiO_2 layer to commercially available mesoporous silicon membranes allowed McCool et al. to produce perfect molecular sieve membranes. They next measured the N_2 gas permeability across these membranes. The mesoporous membrane exhibited aanN_2 permeance of 3.5×10^{-7} $mol/m^2/s/Pa$. Its viscous flow was reduced to less than 1%, while the N_2 permeance decreased to 2.8×10^{-7} $mol/m^2/s/Pa$. During the gas separation investigations, it was found that the pore size was in the intermediate zone between mesopores and micropores. At a temperature of 473 K, the altered membrane exhibited a separation factor of 8.6, indicating a relatively low level of separation between CH_4 and H_2. The membrane was utilized for the challenging separation of p-xylene from o-xylene, with the separation increasing from 1 to 2.1 after modification via ALD. Tran et al. applied a 10-nm thick TiO_2 layer to $\gamma-Al_2O_3$ membranes using ALD, achieving a balance between H_2 permeability and separation properties. At a temperature of 450 K, the permeance of hydrogen was approximately 12.5×10^{-8} $mol/m^2/s/Pa$, and the separation factor for a mixture of hydrogen and carbon dioxide was determined to be 5.8 (Ortner, 2016).

Fu and colleagues have used carbonic anhydrase enzymes contained within 8 nm wide silica mesopores found within AAO templates to produce an incredibly thin enzymatic liquid membrane. During the membrane fabrication process, ALD was employed to perform five cycles of alternate exposures to hexamethyldisilane (HMDS), trimethylchlorosilane (TMCS), and water. The aim was to quantitatively substitute hydrophobic trimethylsilyl groups $Si(CH_3)_3$ for hydrophilic surface silanol groups. This ultra-thin enzymatic liquid membrane demonstrated impressive CO_2 separation capability, with selectivity values of 788 for CO_2/N_2 and 1,500 for CO_2/H_2. Furthermore, it achieved the most elevated combined flux and selectivity yet documented for operation under normal atmospheric circumstances. Carbon-based chemicals are used to generate membrane substrates. Feng et al. utilized ALD to apply zinc oxide (ZnO) onto randomly intertwined multi-walled carbon nanotubes (CNTs). This was done to create textiles that could function as independent membranes. In particular, in the ultrafiltration range, the carbon nanotube (CNT) membranes demonstrated improved water permeability and retention following ALD. In 2004, Chen and colleagues made a pore in a Si_3N_4 thin film using a focused ion beam (FIB). Then, to produce a nanopore with preset dimensions, they used 500 cycles of ALD of Al_2O_3. This method decreased the size of a very large pore to a desired smaller diameter while preserving its original shape. Lastly, it is crucial to remember that by utilizing ALD, it is possible to produce or modify thick membranes. Several research have been undertaken to investigate the expansion of YSZ, a ceramic material with ion conductivity. The objective of these studies is to augment the ionic conductivity of YSZ and reduce the thickness of the electrolyte layer for applications that involve solid oxide fuel cells. To achieve this objective, the use of ALD super cycles involved alternating between ZrO_2 and Y_2O_3 ALD cycles.

Novel activities beyond membrane separation can be achieved by directly depositing a diverse range of materials and structures onto polymeric membranes. This is achieved by carefully choosing the appropriate ALD precursors and deposition techniques. Membranes that have been coated with semiconducting oxides such as ZnO and TiO_2 can be directly endowed with the photocatalytic property. Leveraging ALD, Li et al. successfully fabricated a three-dimensional heterostructure of TiO_2 / ZnO type II on PVDF membranes. As a result, the membranes gained the property of becoming hydrophilic when exposed to light and were able to carry out photocatalysis, as shown in Figure 3.3a. After exposing the sample to light, the water contact angle (WCA) decreased by 82.6% when the most effective ALD sequence was used. This sequence involved one cycle of TiO_2 and three cycles of ZnO. In addition, the flow increased by 33.5%. The TiO_2 / ZnO type II heterostructures, when exposed to visible light, demonstrated the capability to remove around 80% of methylene blue when applied to PVDF membranes. Furthermore, the membranes that were deposited exhibited a significant improvement in fouling resistance. Several factors, including as precursor-substrate interactions, deposition temperature, precursor dosage, and deposition mode, influence the morphology of metal oxides generated during ALD. Zinc oxide was applied onto electrospun nylon nanofibrous membranes of varying morphologies, including ZnO nanoparticles (NPs), highly compacted ZnO NPs, and a 27-nm thick ZnO layer. Therefore, the membranes acquired a range of photocatalytic properties. The dosage and deposition method of the precursor were regulated. All the membranes coated with ZnO were capable of degrading rhodamine B (RhB) when exposed to UV light. Nevertheless, the membranes coated with densely packed ZnO nanoparticles exhibited superior performance. This was a result of the increased surface roughness and surface area of these membranes. ALD exhibits the capacity to not only carry out photocatalytic operations but also enhance electrochemical activity in membranes. Yang et al. employed ALD to enhance the fouling resistance of PVDF membranes by creating conductive layers of aluminum-doped zinc oxide (AZO) (Figure 3.3b). The deposition of a single layer of AZO was achieved by initially depositing 20 cycles of ZnO, followed by the deposition of one cycle of Al_2O_3. A protective layer consisting of a thin 11-nm coating of TiO_2 was applied using ALD onto the surface of the AZO material. This can be observed in the inset of Figure 3.3b. The use of this coating significantly enhanced the endurance of the AZO films and effectively inhibited corrosion in water for a duration of 7 days. Thicker AZO sheets performed better electrochemically even with lower permeability. Bacteria adhered to the membrane with 50 AZO layers 72% less compared to the clean PVDF membrane. Adding 1.5 V of voltage to the AZO50 membranes further decreased bacterial adhesion to 86%. Furthermore, depending on the properties of the deposited materials, it is feasible to obtain other functions such as adsorption. Xiong and colleagues used zinc oxide onto polythene fluoride (PTFE) membranes, thereby imparting the membranes with the capacity to absorb color. Based on the findings of adsorption trials, the PTFE membranes that were applied were effective in eliminating both RhB molecules with positive charge and acid orange molecules with a negative charge from water, demonstrating a reasonable level of efficiency. Despite undergoing many reuse cycles, the deposited PTFE membranes showed no discernible indications of performance degradation. According to Yang et al., Figure 3.3a demonstrates that PVDF membranes possess a significant level of resistance to oils. Experimental tests and molecular dynamic simulations have shown that the membranes coated with TiO_2 and SnO_2 have strongly attached hydration layers on their surfaces. Consequently, the presence of a water cushion on the membranes hindered the interaction between oil and the surface of the membrane. Because water

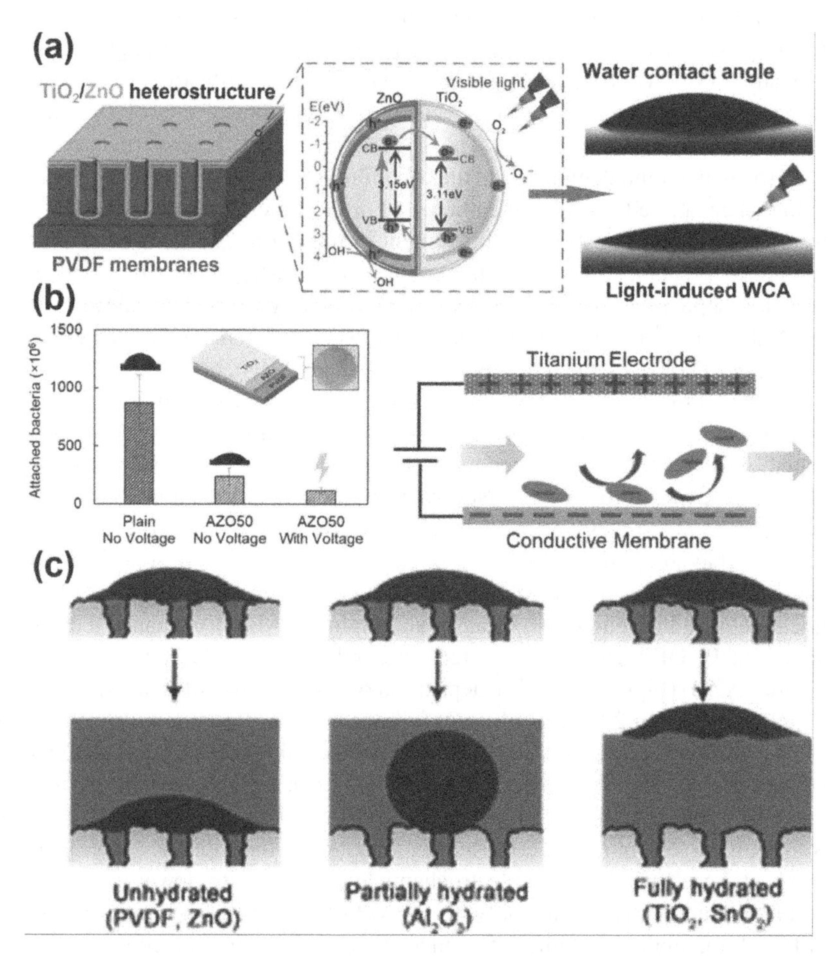

FIGURE 3.3 (a) The photocatalytic mechanism of TiO_2/ZnO type II heterostructures is investigated together with the composition and water-attracting characteristics of PVDF membranes modified using ALD. (b) The application diagram of conductive membranes is examined, together with bacterial attachment to AZO50 membranes at a voltage of −1 V. (c) Possible methods of oil adhesion on the surfaces of PVDF membranes that have not been treated and those that have been coated with different metal oxides by ALD are hypothesized (Xiong et al., 2022).

and oil are immiscible, the fully hydrated membranes acquired an outstanding ability to resist crude oil (Figure 3.3c). In addition, superhydrophilicity can be attained by employing ALD of metal oxides on polymeric substrates with particularly rough surfaces. Li et al. applied titanium dioxide layers onto polypropylene nonwovens, causing the nonwovens to become extremely water-attracting and highly resistant to oil underwater, even though they were originally water-repellent. The polypropylene nonwovens that were deposited exhibited a self-cleaning ability, enabling the organic solvents connected to them to be eliminated only with water washing. This was achieved through the implementation of tailored wettability. Furthermore, the polypropylene nonwovens that were deposited also acquired the chemical stability of the TiO_2 material and could be effectively used in harsh settings. ALD of metal oxides has the capability to transform hydrophilic membranes into hydrophobic membranes, which may seem contradictory. Deposition of a few cycles of Al_2O_3 or TiO_2 improved the surface C–C/C–H bonds of cellulose-based nanopaper membranes. Previously hydrophilic, this treatment turned the membranes hydrophobic. C–C/C–H bonds were found to be formed by either unreacted methyl groups from the organometallic precursors or carbon that was inadvertently adsorbed on the membrane surface after deposition. Furthermore, the modified membranes demonstrated tolerance to high sonication in both polar and aqueous solvents. M-OH terminations on metal oxides have been shown to enhance the hydrogen bonds between cellulose molecules, therefore lowering the hydration effect and weakening the interaction between fibrils and solvents. The altered cellulose membranes were thus stronger. The strong oleophilicity and good hydrophobicity of these membranes suggested that they might find use in oil-water separation. Remarkably, the quantity of oxide deposited, which is based on the number of ALD cycles, can affect the wettability of metal oxides on polymeric substrates. By deposition of Al_2O_3, the thermal and mechanical characteristics of nanofibrous aerogels (NFAs) derived from poly(vinyl alcohol-co-ethylene) (EVOH) were enhanced. The Al_2O_3 layer on the surface and inside the aerogels did not change appreciably after a hundred deposition cycles. The results indicated that ALD had the ability to apply consistent layers onto highly porous surfaces. The aerogels exhibited hydrophobic behavior when exposed to fewer than eight ALD cycles, and hydrophilic behavior when exposed to more than eight ALD cycles. The change in wettability was influenced by various factors, such as structural and chemical elements. When the number of ALD cycles was less than seven, there was a tendency for the Al_2O_3 to be deposited on the fibrils in the form of little protrusions. The formation of hierarchical structures was achieved through the combination of coarse fibrils with very porous NFAs. These structures could store a greater amount of air and enhance the hydrophobic properties of the EVOH NFAs as they were being built. By increasing the number of ALD cycles, the C–OH bonds on NFAs transformed hydrophilic Al_2O_3. Additionally, the hierarchical structures gradually disintegrated due to the development of homogenous Al_2O_3 deposition layers. Consequently, the WCAs decreased in proportion to the increase in the number of ALD cycles. The aerogels that underwent six cycles of ALD exhibited the greatest degree of hydrophobicity, while the aerogels that underwent eight cycles of ALD experienced a change in their capacity to repel water. The aerogels deposited utilizing a six-cycle method demonstrated exceptional flexibility and remarkable elastic recovery. Even after undergoing 500 cycles of stress-strain tests, they were able to retain 70% of their original maximum stress and Young's modulus. Considering the exceptional mechanical stability and strong hydrophobic properties, the aerogels produced using a six-cycle technique were discovered to be highly effective in absorbing oil. The aerogels exhibited absorption capacities ranging from 31 to 73 g/g for different organic solvents, and they also showed outstanding reusability. Moreover, it is crucial to underscore that this study does not directly address membranes. Conversely, the technique of modifying the wettability by employing ALD cycles can be readily extended to fabricate functional membranes with switchable wettability (Xiong et al., 2022).

Xu and colleagues used Al_2O_3 onto PTFE membranes and subsequently successfully made the membranes hydrophilic. The growth of Al_2O_3 granules on the surface of the PTFE membranes was observed to increase with the number of ALD cycles. Furthermore, the membranes' constituent fibrils progressively increased in diameter. A substantial decrease in the effective pore size resulted from the fibrils being completely coated with Al_2O_3 via specialized ALD. WCA of PTFE membranes dropped from about 130° to less than 20° after 500 cycles of ALD deposition. Thanks to their hydrophilized surface and reduced pore size, the modified PTFE membranes increased pure water flux (PWF) by 67.7% and retained 190-nm polystyrene (PS) nanospheres by 11.4%. This work has shown that ALD can successfully overcome the trade-off effect that has long been associated with polymeric membranes. The results of this study were confirmed by the actions of other researchers who applied different substances onto distinct membranes. Based on the results shown in Figure 3.4, PVDF membranes coated with TiO_2 showed a gradual reduction in WCAs. At the same time, they also showed enhancements in selectivity and permeability. Hollow-fiber membranes that have undergone the necessary ALD treatment have demonstrated concurrent improvements in both water permeability and rejection. The upcoming part will address this topic. In addition, inorganic membranes can also be influenced by the trade-off effect, which can be mitigated by adopting absorption-limited deposition (ALD) with suitable materials.

3.5 ELECTROLESS PORE-PLATING (ELP-PP)

The objective of the ELP-PP technique is to seamlessly incorporate palladium into the pores, so preserving the essential benefits of ELP while simultaneously providing specific benefits that are associated with the technology that is now in use. Within the ELP process, the first step involves the integration of particles onto the membrane's surface. They can attract other particles, which are then loaded together until they construct a bridge that gradually develops throughout the process. After that, the palladium nanoparticles deposit onto the bridge until the entire pores are covered because of the increase in the thickness of the film. Therefore, the thickness of the Pd membrane is determined by the pore size of the support. The size of the pores on the support surface is approximately three times less than the thickness of the file, which is typically three times larger (Alique et al., 2020). On the other hand, the fact that the ELP process is an auto-catalyst, in addition to the fact that palladium nanoparticles are deposited on the surface area of the membrane, makes it difficult to produce a thin film. To find a solution to this issue, the ELP-PP approach was examined to minimize

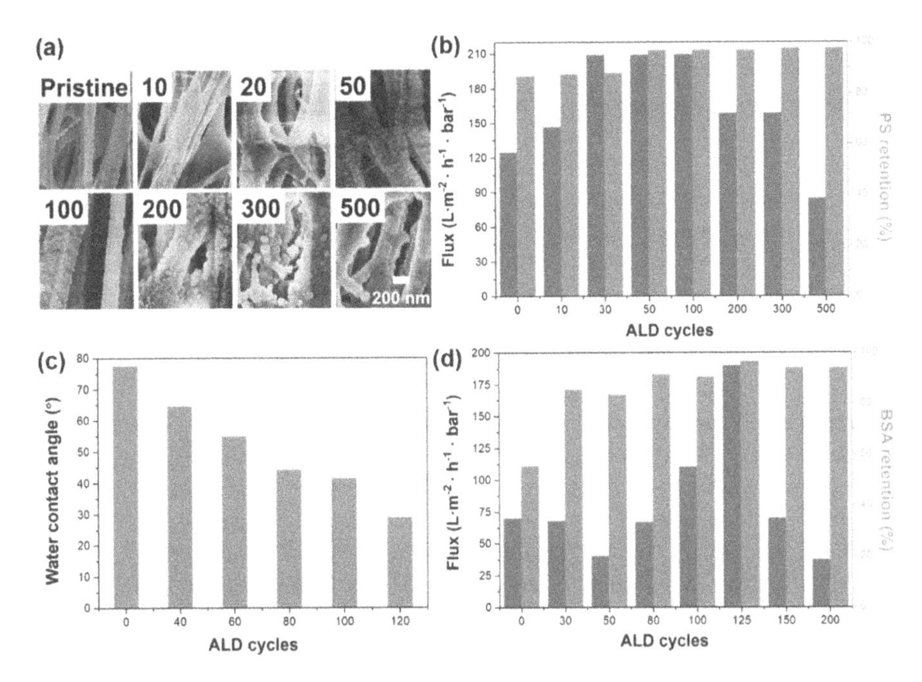

FIGURE 3.4 Selectivity and permeability can be traded off using atomic layer deposition (ALD) of metal oxides on polymeric membranes. (a) Scanning electron microscopy (SEM) pictures of the PTFE membranes with various atomic layer deposition (ALD) cycles coated with Al_2O_3. (b) 190-nm polystyrene (PS) nanosphere retention and permeability water flux (PWF) by PTFE membranes treated with various ALD cycles of Al_2O_3 deposition. Angles of contact with water measured (c). (d) Bovine serum albumin (BSA) retention and permeate water flow (PWF) by PVDF membranes treated with various ALD cycles of TiO_2 deposition (Xiong et al., 2022).

the flaws that were formed on the membrane. Incorporating palladium nanoparticles into the holes of the support material and constructing a dense membrane are the goals of this technology. One of the most acceptable methods is the inclusion of a thin film of palladium (or alloys) over a porous support using electroless plating. This method is advantageous due to the uniformity and hardness of deposits, the simplicity of the procedure, and the superior permeability of the material. On the other hand, one of the most significant disadvantages associated with the utilization of these membranes on a wide scale is the elevated investment cost. A significant portion of this increased cost can be attributed to the palladium and support prices, in addition to the quantity of membranes that were rejected throughout the material manufacturing process. The fabrication of ultra-thin layers of palladium, the reduction of the amount of metal, and the maximization of the membrane's permeability capability are the primary strategies that are utilized to save money. There are a variety of alternatives that have been suggested in the research literature. Among these options are those by Zhang et al. that use a vacuum to fill the support's pores with palladium; Souleimanova et al. that combine electroless plating with osmosis to facilitate the incorporation of Pd around the pore area; and Yogo et al. that prepare pore-filled membranes in which the palladium is incorporated as an internal layer on the support cross-section. All the same, flaws are not unusual during the synthesis of these incredibly thin films. These defects include wrinkles, cracks, and pinholes. Considering this, fixing these problems might be considered as a workable strategy to lower the total cost of membrane preparation. Li and colleagues developed an electroless plating and osmosis technique that guarantees the total elimination of Pd layer flaws without sacrificing permeability or raising the thickness of the metal. Zeng and colleagues used a defect-sealing method known as point plating to introduce palladium salt and the reducing agent from opposing sides of the membrane at the same time. Lu and colleagues propose another approach that includes intermetallic diffusion bonding into the repair procedure. Antonio and colleagues have put up a new method called electroless pore-plating (ELP-PP) to reduce the occurrence of defects. This method produces dense membranes in a single specialized step by forming a solid stainless steel support, which can minimizes the leakage of any fluid by blocking the pores or gaps present on membrane. The Pd source and the reducing agent are guaranteed to interact inside the pores of the support by introducing both reactants from opposing sides of the support. Even though this technology has shown that it is possible to generate composite membranes that are fully free of defects, the conditions that are utilized throughout the synthesis phase are still being investigated. On the other hand, not only has the fabrication of membranes garnered an increasing amount of attention, but the modeling of hydrogen permeation through these Pd-based membranes has also been a subject of significant interest over the past several years. ELP-PP was used to construct composite Pd-porous stainless steel (Pd–PSS) membranes, and Antonio and his colleagues investigated the effect that the concentration of the reducing agent had on the process. It has been demonstrated through the findings that this variable influences the shape of the Pd layer as well as the permeability of the membrane. For modeling the permeation outcomes, a rigorous technique has been utilized, which is founded on the precise specification of the individual transport procedure phases (Martinez-Diaz et al., 2021).

The permeation behavior of numerous Pd-composite membranes was investigated by Alique et al., who used both traditional ELP and a unique pore-plating approach (ELP-PP) to manufacture the membranes. The membranes were then placed on ceramic and metallic supports. The membrane that was created using the conventional electroluminescence (ELP) technique over a tubular ceramic support exhibits a permeability that falls within the range of (2.5–3.6)10^{-6} mol s^{-1} $bar^{-0.5}$ m. Additionally, it possesses approximately perfect ideal selectivity and roughly 14 m of Pd thickness. With the alternate preparation process, ELP-PP, it was observed that the hydrogen fluxes were lower, with permeability ranging from (5.8–38.5)10^{-7} mol s^{-1} $bar^{-0.5}$ m. This was the case even though the Pd thickness was reduced to 8 m. Additionally, perfect selectivity was also observed. To explain this behavior, it is necessary to consider the fact that the pore-plating process results in the deposition of platinum not only on the surface of the support but also within the pores. This results in an effective thickness of platinum that is greater than what is obtained by the standard ELP method. Additionally, it has been discovered that in this instance, there is a reduced deposition of palladium within the pores of the support, which results in a lower resistance to the hydrogen permeation that occurs in ceramic-supported membranes. Using PSS supports, Alberto and his colleagues were able to successfully construct dense Pd layers with a thickness of around 17 μm. It has been demonstrated that the addition of an intermediate graphite layer before the deposition of Pd results in an improvement in the membrane's overall performance as effectively as its durability. During the operating temperatures ranging from 350°C to 450°C, these membranes have a remarkable hydrogen permeance of 3.24×10^{-4} to 4.33×10^{-4} mol/m^2/s/Pa$^{0.5}$. Furthermore, they exhibit an exceptional selectivity of H_2/N_2 that is equal to or greater than 10,000. To simulate the H_2/N_2 separation process in membrane permeators, computational fluid dynamics (CFD) models have been utilized. The prediction of hydrogen flux through the membrane is accomplished by these models through the utilization of species transport equations and Sieverts' law. Indicating that the concentration polarization near the membrane surface does not considerably restrict hydrogen permeability, simulations have shown strong agreement with experimental results, indicating that this is the case. It is important to highlight that the Pd membranes have a remarkable mechanical robustness. Tests that involved the reversal of permeation direction (from in-out to out-in) indicated that the membranes retained their performance, which suggests that they have strong mechanical stability and tolerance to stress. This is an essential characteristic for practical applications, which face the possibility of encountering a wide range of operational situations. The performance of the membrane has been investigated in these studies under a variety of conditions, including varying feed pressures and hydrogen concentrations. The membranes are capable of handling binary gas mixtures and maintaining high hydrogen permeance even when nitrogen is present. However, a drop in performance was found due to concentration-polarization events at greater nitrogen concentrations. These investigations have shed light on the membranes' capabilities.

3.6 CONCLUSION

In this book, the many approaches that have been used to build membranes that are both effective and durable for the purpose of hydrogen separation or purification are discussed. Electroless plating, CVD, and PVD are some of the fabrication techniques that are highlighted in this chapter. Each approach comes with its own set of benefits and difficulties, and it is adapted to meet the requirements of applications and performance standards. Electroless pore-plating, also known as ELP-PP, is a process that has the potential to manufacture dense and defect-free palladium (Pd) layers on porous substrates. This makes it a particularly attractive approach. The mechanical strength, permeability, and selectivity of the membranes can be greatly improved through the utilization of techniques such as the incorporation of intermediate layers (e.g., graphite) and the utilization of modern deposition procedures. The implementation of these enhancements is important to achieve the high levels of hydrogen purity that are required for industrial applications. Together, the CFD modeling and experimental approach guarantee that the membranes that have been produced match the specified performance standards, which makes it easier to put them into practice. The industrial significance of hydrogen separation membranes is highlighted by the developments in fabrication techniques that are covered in this chapter. The purification of hydrogen is an essential stage in a variety of clean energy processes, including the manufacture and storage of hydrogen fuel. These membranes provide a viable solution for the purification of hydrogen. This chapter emphasizes the significance of continuing research and development to improve the effectiveness and cost-effectiveness of these technologies.

REFERENCES

Alique, D., Martinez-Diaz, D., Sanz, R., Calles, J.A., 2018. Review of supported Pd-based membranes preparation by electroless plating for ultra-pure hydrogen production. *Membranes (Basel)* 8, 5.

Alique, D., Sanz, R., Calles, J.A., 2020. Pd membranes by electroless pore-plating: synthesis and permeation behavior. In *Current Trends and Future Developments on (Bio-) Membranes*; Basile, A., Ed. Elsevier, pp. 31–62.

Böke, F., Giner, I., Keller, A., Grundmeier, G., Fischer, H., 2016. Plasma-enhanced chemical vapor deposition (PE-CVD) yields better hydrolytical stability of biocompatible SiOx thin films on implant alumina ceramics compared to rapid thermal evaporation physical vapor deposition (PVD). *ACS Appl Mater Interfaces* 8, 17805–17816. https://doi.org/10.1021/acsami.6b04421

Guo, Y., Zou, H., Wu, H., Zhou, L., Liu, H., Zhang, X., 2014. Preparation of palladium membrane by bio-membrane assisted electroless plating for hydrogen separation. *Int J Hydrogen Energy* 39, 7069–7076.

Habib, M.A., Harale, A., Paglieri, S., Alrashed, F.S., Al-Sayoud, A., Rao, M.V., Nemitallah, M.A., Hossain, S., Hussien, M., Ali, A., others, 2021. Palladium-alloy membrane reactors for fuel reforming and hydrogen production: a review. *Energy Fuels* 35, 5558–5593.

Itoh, N., Akiha, T., Sato, T., 2005. Preparation of thin palladium composite membrane tube by a CVD technique and its hydrogen permselectivity. *Catal Today* 104, 231–237.

Khatib, S.J., Oyama, S.T., 2013. Silica membranes for hydrogen separation prepared by chemical vapor deposition (CVD). *Sep Purif Technol* 111, 20–42.

Kudapa, V.K., Paliyal, P.S., Mondal, A., Mondal, S., 2024. A critical review of fabrication strategies, separation techniques, challenges, and future prospects for the hydrogen separation membrane. *Fusion Sci Technol* 1–23. https://doi.org/10.1080/15361055.2023.2290898

Liu, G., Li, P., Zhang, X., Zhang, L., Yang, T., 2019. Development of hydrogen-selective membranes using chemical vapor deposition (CVD) techniques. *Journal of Membrane Science*, 582, 156–165.

Ma, M., Mao, Y., Gupta, M., Gleason, K.K., Rutledge, G.C., 2005. Superhydrophobic fabrics produced by electrospinning and chemical vapor deposition. *Macro-Molecules* 38, 9742–9748.

Martinez-Diaz, D., Leo, P., Sanz, R., Carrero, A., Calles, J.A., Alique, D., 2021. Life cycle assessment of H2-selective Pd membranes fabricated by electroless pore-plating. *J Clean Prod* 316, 128229.

Omidifar, M., Babaluo, A.A., Jamshidi, S., 2024. H2 permeance and surface characterization of a thin (2 $μ$m) Pd-Ni composite membrane prepared by electroless plating. *Chem Eng Sci* 283, 119370.

Ortner, K., 2016. Palladium gas separation membranes for hydrogen separation. *Materials*, 9(5), 359.

Pal, N., Agarwal, M., Maheshwari, K., Solanki, Y.S., 2020. A review on types, fabrication and support material of hydrogen separation membrane. *Mater Today Proc* 28, 1386–1391.

Wang, Y., Chen, H., Li, J., Zhang, W., Liu, X., 2020. Evaluation of catalyst effects on hydrogen membrane fabrication via chemical vapor deposition (CVD): A comparison of nickel, palladium, and platinum. *Journal of Membrane Science*, 611, 118–273.

Weber, M., Drobek, M., Rebière, B., Charmette, C., Cartier, J., Julbe, A., Bechelany, M., 2020. Hydrogen selective palladium-alumina composite membranes prepared by atomic layer deposition. *J Memb Sci* 596, 117701.

Xiong, S., Qian, X., Zhong, Z., Wang, Y., 2022. Atomic layer deposition for membrane modification, functionalization and preparation: a review. *J Memb Sci* 658, 120740.

Yamazaki, Y., Kondo, T., Nakagawa, T., Matsumoto, H. 2017. Evaluation of deposition temperature and precursor gas content on the properties of hydrogen-selective membranes fabricated via chemical vapor deposition (CVD). *Journal of Membrane Science*, 536, 120–128.

Zhang, H., Wang, Y., Li, Q., Sun, J., 2018. Investigation of carbon-based nanomaterials as substrates for hydrogen membrane fabrication using chemical vapor deposition (CVD). *International Journal of Hydrogen Energy*, 43(30), 14187–14196.

4 Challenges, Applications, and Performance Evaluation of Hydrogen Separation Membranes

4.1 APPLICATIONS OF HYDROGEN MEMBRANE

4.1.1 Hydrogen Separation

Hydrogen membranes play a crucial role in separating hydrogen from gas mixtures, enabling the extraction of high-purity hydrogen (Amin et al., 2023). This application is vital in various industries, including petrochemical and refining processes, where pure hydrogen is required for specific reactions and applications (Dube et al., 2023; Singla et al., 2022).

Table 4.1 provides a snapshot of the diverse applications where hydrogen separation membranes play a crucial role. Here are real-world case studies, practical examples, and the specific membrane technologies employed:

Case Study 1: Hydrogen Separation in Refining Processes (Faraji et al., 2005)

- **Industry:** Petrochemical Refining
- **Challenge:** Refineries often require pure hydrogen for catalytic processes. Traditional separation methods, such as pressure swing adsorption (PSA) or cryogenic distillation, are energy-intensive and costly.
- **Solution:** Polymeric hydrogen membranes have been employed in refining processes to selectively separate hydrogen from gas mixtures. A case study involves the use of polyimide membranes that exhibit high hydrogen permeability, allowing for efficient separation at lower operating costs.
- **Outcome:** Improved energy efficiency, reduced operating costs, and increased overall productivity in the refining process.

Case Study 2: On-site Hydrogen Production for Electronics Manufacturing (Cardona et al., 2023)

- **Industry:** Electronics Manufacturing
- **Challenge:** Electronics manufacturing requires ultra-pure hydrogen for processes like annealing and sintering. Transporting hydrogen from external sources poses logistical challenges and potential impurity risks.

DOI: 10.1201/9781003590682-4

TABLE 4.1
Sample Table Outlining Various Applications of Hydrogen Separation Membranes

Application	Description
Hydrogen production	Membranes are used in steam reforming, water-gas shift reactions, and other processes to separate hydrogen.
Hydrogen purification	Removal of impurities such as CO, CO_2, and H_2S from hydrogen streams to meet purity requirements.
Petrochemical industries	Hydrogen membranes aid in processes like ammonia synthesis, methanol production, and hydrocracking.
Hydrogen fuel cells	Membranes enable the separation of hydrogen from reactant streams, improving the efficiency of fuel cells.
Renewable energy storage	Membranes facilitate the purification and storage of hydrogen produced from renewable sources like solar and wind energy.
Hydrogen recovery	Membranes are utilized in processes such as refinery off-gas recovery and byproduct hydrogen streams.
Electronics manufacturing	Hydrogen is used in semiconductor manufacturing processes, and membranes help ensure its purity.
Aerospace applications	Hydrogen membranes aid in fuel purification and storage for hydrogen-powered aircraft and rockets.
Chemical synthesis	Membranes are employed in various chemical reactions where hydrogen separation is required.

Source: Dube et al. (2023).

- **Solution:** Ceramic hydrogen membranes have been deployed on-site for hydrogen separation directly within electronics manufacturing facilities. These membranes effectively remove impurities, ensuring a constant supply of high-purity hydrogen.
- **Outcome:** Enhanced process reliability, reduced dependence on external hydrogen suppliers, and improved product quality in electronics manufacturing.

Case Study 3: Hydrogen Recovery in Ammonia Production (Seiiedhoseiny et al., 2024)

- **Industry:** Chemical Manufacturing (Ammonia Production)
- **Challenge:** Ammonia synthesis involves the production of hydrogen as a byproduct. Efficient recovery of hydrogen from the synthesis gas stream is crucial for minimizing waste and optimizing resource utilization.
- **Solution:** Metal membrane technologies, such as palladium-based membranes, have been implemented for hydrogen recovery in ammonia production. These membranes selectively permeate hydrogen, allowing for its extraction from the synthesis gas.
- **Outcome:** Reduced waste, increased ammonia production efficiency, and a more sustainable and resource-efficient manufacturing process.

Practical Example: Industrial Hydrogen Purification Plant

- **Industry:** Industrial Gas Manufacturing
- **Scenario:** An industrial gas manufacturing plant relies on hydrogen as a key product. To meet the stringent purity requirements of customers, a hydrogen purification plant is established.
- **Technology:** Composite membranes combining metallic and polymeric components are employed for enhanced selectivity and durability. These membranes effectively separate impurities, providing ultra-pure hydrogen for industrial use.
- **Result:** Consistent delivery of high-purity hydrogen to customers, meeting strict quality standards and establishing a reputation for reliability in the industrial gas market.

Emerging Technology: Advanced Metal Membranes

- **Innovation:** Ongoing research focuses on the development of advanced metal membranes, such as alloys with improved hydrogen permeability and resistance to impurities.
- **Potential Benefits:** These innovations aim to further enhance the efficiency of hydrogen separation processes, reduce operational costs, and broaden the applicability of membrane technology across various industries.

In each of these examples, the specific membrane technologies employed showcase the versatility and adaptability of hydrogen membranes in addressing the unique challenges of different industries, ultimately contributing to increased efficiency, cost-effectiveness, and sustainability.

4.1.2 Hydrogen Purification

In industries such as electronics and metallurgy, where stringent purity requirements exist, hydrogen membranes are employed for the purification of hydrogen gas. The membranes selectively remove impurities, ensuring the production of ultra-pure hydrogen for critical processes (Bernardo et al., 2020).

Table 4.2 highlights various critical aspects of hydrogen purification, including challenges and potential solutions or advancements in each area.

Here are real-world case studies, practical examples, and specific membrane technologies employed for optimal results:

Case Study 1: Electronics Industry—Semiconductor Manufacturing

- **Industry:** Electronics (Semiconductor Manufacturing)
- **Challenge:** Semiconductor manufacturing demands ultra-pure hydrogen for processes like silicon wafer manufacturing. Traditional purification methods might not meet the stringent purity requirements efficiently.
- **Solution:** Palladium-based membranes have been implemented for hydrogen purification in semiconductor plants. These membranes selectively allow hydrogen to pass through, ensuring a high level of purity.

TABLE 4.2
Different Aspects of Hydrogen Purification

Aspect	Description	Challenges	Solutions/ Advancements
Membrane material	Types of materials used in hydrogen separation membranes, including polymers, metals, and ceramics.	Limited selectivity and stability of materials.	Development of novel materials with enhanced properties.
Purity requirements	Standards and specifications for hydrogen purity levels in various applications.	Meeting stringent purity requirements economically.	Optimization of membrane design and system integration.
Operating conditions	Temperature, pressure, and feed gas composition requirements for membrane operation.	Limited operability under extreme conditions.	Research on adaptive membranes and advanced coatings.
Cost	Economic considerations including membrane fabrication, installation, and maintenance costs.	High initial investment and operational expenses.	Advancements in manufacturing techniques and scale-up.
Scalability	Ability to scale up membrane systems to meet industrial-scale hydrogen purification needs.	Limited scalability of membrane technologies.	Modular system designs and integration with existing infrastructure.
Reliability	Dependability and longevity of membranes under continuous operation in industrial settings.	Membrane degradation and fouling over time.	Development of robust membrane architectures and cleaning protocols.
Environmental impact	Assessment of the environmental footprint associated with membrane production and operation.	Energy consumption, waste generation, and emissions.	Pursuit of sustainable materials and energy-efficient processes.

Source: Bernardo et al. (2020).

- **Outcome:** Improved yield in semiconductor production, enhanced reliability of processes, and reduced contamination risks, contributing to the production of high-quality electronic components.

Case Study 2: Energy Sector—Hydrogen Fuel Production

- **Industry:** Renewable Energy (Hydrogen Fuel Production)
- **Challenge:** Renewable energy projects often involve electrolysis to produce hydrogen from water. Ensuring the purity of hydrogen generated through this process is crucial for fuel cell applications.

- **Solution:** Alumina-based ceramic membranes have been utilized for purifying hydrogen produced via electrolysis. These membranes effectively remove impurities, resulting in high-purity hydrogen suitable for fuel cell applications.
- **Outcome:** Increased efficiency in renewable hydrogen production, supporting the growth of clean energy initiatives and hydrogen as a sustainable fuel source.

Case Study 3: Chemical Industry—Specialty Chemicals Manufacturing

- **Industry:** Specialty Chemicals Manufacturing
- **Challenge:** Specialty chemicals manufacturing requires precise control over the composition of reactants. Contaminants in hydrogen can negatively impact product quality.
- **Solution:** Composite membranes, combining polymer and metallic components, are employed for hydrogen purification in specialty chemical plants. These membranes offer a balance of selectivity and durability.
- **Outcome:** Consistent production of high-quality specialty chemicals, reduced waste, and improved overall efficiency in the chemical manufacturing process.

Practical Example: Hydrogen Refueling Station

- **Industry:** Transportation (Hydrogen Fueling Station)
- **Scenario:** A hydrogen refueling station supplying fuel for fuel cell vehicles requires a reliable source of high-purity hydrogen to meet vehicle fueling standards.
- **Technology:** Polymeric membranes with high hydrogen selectivity are utilized in the purification process at the refueling station. These membranes efficiently remove impurities, ensuring the delivery of pure hydrogen to fuel cell vehicles.
- **Result:** Reliable operation of the refueling station, meeting stringent purity standards for fuel cell vehicles, and contributing to the growth of hydrogen as a clean transportation fuel.

Emerging Technology: Mixed Matrix Membranes

- **Innovation:** Ongoing research explores the development of mixed matrix membranes, combining traditional polymeric materials with advanced nanoparticle fillers.
- **Potential Benefits:** These innovative membranes aim to enhance selectivity, permeability, and durability, offering improved performance in hydrogen purification applications across various industries.

In these examples, the specific membrane technologies employed demonstrate the adaptability of hydrogen purification methods to diverse industrial requirements, contributing to increased efficiency, product quality, and sustainability.

4.1.3 Hydrogen Storage

Hydrogen membrane technology is instrumental in creating efficient and compact storage solutions for hydrogen. Membranes facilitate the storage of high-purity hydrogen, making it a versatile energy carrier for various applications, including fuel cells and transportation (Barthélémy et al., 2017).

Here are real-world case studies, practical examples, and specific membrane technologies employed for optimal results in hydrogen storage:

Case Study 1: Automotive Industry—Hydrogen Fuel Cell Vehicles

- **Industry:** Automotive (Hydrogen Fuel Cell Vehicles).
- **Challenge:** Hydrogen fuel cell vehicles require efficient and safe storage solutions for hydrogen to enable extended driving ranges and quick refueling.
- **Solution:** Metal hydride storage systems, employing membranes for controlled hydrogen release, have been integrated into fuel cell vehicles. These membranes allow the release of hydrogen on-demand for power generation.
- **Outcome:** Enhanced safety, increased energy density, and improved driving range for hydrogen fuel cell vehicles.

Case Study 2: Renewable Energy Storage—Power-to-Gas

- **Industry:** Renewable Energy (Power-to-Gas).
- **Challenge:** Power-to-gas projects aim to store excess renewable energy by converting it into hydrogen. Effective and compact hydrogen storage solutions are crucial for grid balancing.
- **Solution:** Advanced porous materials like metal-organic frameworks (MOFs) are employed as membranes for hydrogen adsorption-based storage systems. These membranes enable efficient and reversible hydrogen storage.
- **Outcome:** Improved energy storage efficiency, reduced reliance on conventional power sources, and increased flexibility in managing renewable energy fluctuations.

Case Study 3: Industrial Manufacturing—Metal Hydride Storage

- **Industry:** Metal Manufacturing.
- **Challenge:** Certain industrial processes require a controlled and on-demand supply of hydrogen. Traditional storage methods may not provide the required flexibility and safety.
- **Solution:** Metal hydride storage systems with membranes for controlled hydrogen release have been implemented in metal manufacturing plants. These membranes regulate the release of hydrogen for specific applications.
- **Outcome:** Increased process efficiency, enhanced safety, and improved control over hydrogen usage in metal manufacturing processes.

Practical Example: On-site Hydrogen Storage for a Microgrid

- **Industry:** Microgrid Energy System.
- **Scenario:** A microgrid powered by a combination of renewable energy and conventional sources requires a reliable and flexible energy storage solution for intermittent renewable energy generation.
- **Technology:** Polymer-based membranes are employed in a solid-state hydrogen storage system. These membranes enable the safe storage and controlled release of hydrogen for power generation during periods of low renewable energy availability.
- **Result:** Increased resilience of the microgrid, reduced dependence on the main power grid, and improved sustainability through effective hydrogen storage.

Emerging Technology: Liquid Organic Hydrogen Carriers (LOHC)

- **Innovation:** Research is ongoing in the development of liquid organic hydrogen carriers (LOHC) as an alternative hydrogen storage method. Membrane technologies are explored to enhance the efficiency of LOHC systems.
- **Potential Benefits:** LOHC systems have the potential to provide a safe, energy-dense, and reversible hydrogen storage solution, offering advantages in terms of transport, distribution, and energy system integration.

In these examples, the specific membrane technologies employed highlight the diversity of hydrogen storage solutions across different industries, showcasing advancements that contribute to increased safety, efficiency, and applicability of hydrogen as an energy storage medium.

4.1.4 Hydrogen Recovery

Industries generating hydrogen as a byproduct can utilize membranes for the recovery of hydrogen from process streams. This not only ensures the efficient utilization of resources but also contributes to the overall sustainability of industrial processes (Moral et al., 2024).

Hydrogen recovery is a crucial application, especially in industries where hydrogen is produced as a byproduct. Here are real-world case studies, practical examples, and specific membrane technologies employed for optimal results in hydrogen recovery:
Case Study 1: Ammonia Production—Hydrogen Byproduct Recovery

- **Industry:** Chemical Manufacturing (Ammonia Production).
- **Challenge:** Ammonia synthesis generates hydrogen as a byproduct. Efficient recovery of hydrogen is essential for optimizing resource utilization and minimizing waste.
- **Solution:** Palladium-based membranes have been implemented for selective hydrogen recovery in ammonia production plants. These membranes allow the separation and extraction of hydrogen from the synthesis gas stream.

- **Outcome:** Reduced waste, increased ammonia production efficiency, and a more sustainable and resource-efficient manufacturing process.

Case Study 2: Petrochemical Industry—Refinery Hydrogen Recovery

- **Industry:** Petrochemical Refining.
- **Challenge:** Refineries often produce hydrogen as a byproduct of various processes. Efficient recovery is necessary for ensuring the availability of high-purity hydrogen for catalytic reactions.
- **Solution:** Ceramic membranes with high hydrogen selectivity are employed for the recovery of hydrogen in petrochemical refineries. These membranes aid in the separation and purification of hydrogen from other gases.
- **Outcome:** Enhanced efficiency in refining processes, reduced environmental impact, and increased utilization of hydrogen as a valuable resource.

Case Study 3: Biogas Upgrading—Hydrogen Extraction

- **Industry:** Renewable Energy (Biogas Upgrading).
- **Challenge:** Biogas produced from organic waste contains a significant amount of hydrogen. Efficient extraction of hydrogen from biogas is essential for maximizing energy output.
- **Solution:** Polymeric membranes with high hydrogen permeability are utilized for the separation and recovery of hydrogen from biogas. These membranes selectively allow the passage of hydrogen, leaving behind other gases.
- **Outcome:** Increased energy yield from biogas, improved environmental sustainability, and the utilization of organic waste for hydrogen production.

Practical Example: Industrial Hydrogen Byproduct Recovery System

- **Industry:** Various Industrial Processes
- **Scenario:** Industries generating hydrogen as a byproduct in different processes implement a centralized recovery system.
- **Technology:** Composite membranes with a combination of metallic and polymeric components are employed for their selectivity and durability. These membranes efficiently recover hydrogen from various industrial process streams.
- **Result:** Optimized resource utilization, reduced waste, and the establishment of a sustainable and cost-effective hydrogen recovery system across multiple industrial applications.

Emerging Technology: Membrane Distillation for Hydrogen Recovery

- **Innovation:** Membrane distillation, a novel technique combining thermal and membrane processes, is being explored for hydrogen recovery.
- **Potential Benefits:** This innovation aims to provide an energy-efficient and scalable method for recovering hydrogen from various gas mixtures, further enhancing the economic viability of hydrogen recovery systems.

In these examples, the specific membrane technologies employed showcase the adaptability of hydrogen recovery methods to diverse industrial settings, contributing to increased efficiency, sustainability, and the responsible use of hydrogen as a valuable resource.

4.1.5 Hydrogen Sensors

Hydrogen membranes are integrated into sensor technologies for detecting and monitoring hydrogen levels in diverse environments. This application is critical for safety considerations in industries where hydrogen is handled, stored, or produced (Gamboa and Fernandes, 2024).

Hydrogen sensors play a crucial role in detecting and monitoring hydrogen levels in various environments for safety and control. Here are real-world case studies, practical examples, and specific membrane technologies employed for optimal results in hydrogen sensors:

Case Study 1: Industrial Hydrogen Monitoring System

- **Industry:** Chemical Manufacturing.
- **Challenge:** Chemical manufacturing processes involving hydrogen present potential safety risks. Reliable and real-time monitoring of hydrogen levels is essential for maintaining a safe working environment.
- **Solution:** Solid-state hydrogen sensors utilizing thin-film palladium membranes are deployed in key areas of the manufacturing plant. These sensors react to changes in hydrogen concentration, providing rapid and accurate feedback.
- **Outcome:** Improved safety measures, quick response to potential leaks, and enhanced overall operational safety in the chemical manufacturing facility.

Case Study 2: Hydrogen Fueling Station Safety

- **Industry:** Hydrogen Fueling Infrastructure.
- **Challenge:** Hydrogen fueling stations require continuous monitoring to ensure safe handling and prevent potential hazards during fueling operations.
- **Solution:** Fiber optic sensors with palladium-coated membranes are integrated into the infrastructure of hydrogen fueling stations. These sensors provide real-time data on hydrogen concentrations, allowing for immediate adjustments if needed.
- **Outcome:** Enhanced safety protocols, prevention of potential accidents, and increased confidence in the use of hydrogen fueling stations for fuel cell vehicles.

Case Study 3: Laboratory Hydrogen Detection

- **Industry:** Research and Development Laboratories.
- **Challenge:** Laboratories working with hydrogen gas need precise detection methods to ensure the safety of researchers and prevent accidental releases.

- **Solution:** Nanostructured composite membranes with polymers and metal oxides are integrated into compact hydrogen sensors for laboratory use. These sensors offer high sensitivity and quick response times.
- **Outcome:** Improved safety in laboratory settings, enabling accurate detection of hydrogen leaks and facilitating research involving hydrogen gas.

Practical Example: Warehouse Hydrogen Monitoring

- **Industry:** Warehouse and Storage Facilities.
- **Scenario:** Warehouses storing hydrogen-containing materials or using hydrogen fuel cells for equipment need monitoring systems to ensure the safety of workers and the stored goods.
- **Technology:** Wireless hydrogen sensors with robust polymer membranes are strategically placed throughout the warehouse. These sensors continuously monitor hydrogen levels and transmit data to a centralized control system.
- **Result:** Early detection of leaks, improved worker safety, and the prevention of potential hazards in hydrogen-utilizing warehouse environments.

Emerging Technology: Graphene-Based Hydrogen Sensors

- **Innovation:** Ongoing research explores the use of graphene-based sensors for hydrogen detection. Graphene's unique properties offer the potential for highly sensitive and selective hydrogen sensing.
- **Potential Benefits:** If successfully implemented, graphene-based sensors could provide even higher sensitivity, faster response times, and increased durability for hydrogen detection applications.

In these examples, the specific membrane technologies employed illustrate the diverse approaches to hydrogen sensing, emphasizing the importance of safety in various industrial and research settings. Advancements in sensor technologies contribute to more efficient and reliable hydrogen detection systems, fostering the safe use and integration of hydrogen in different environments.

4.1.6 Hydrogen Fueling Stations

The development of hydrogen fueling infrastructure relies on efficient hydrogen separation and purification. Membranes are employed in fueling stations to ensure the delivery of high-purity hydrogen for fuel cell vehicles, contributing to the growth of the hydrogen transportation sector (Isaac and Saha, 2023).

Hydrogen fueling stations are crucial components of the hydrogen infrastructure, supporting the growth of fuel cell vehicles. Here are real-world case studies, practical examples, and specific membrane technologies employed for optimal results in hydrogen fueling stations:

Case Study 1: Public Hydrogen Fueling Station Deployment

- **Location:** California, USA.
- **Challenge:** To promote the adoption of hydrogen fuel cell vehicles, a network of public hydrogen fueling stations was needed to provide convenient access for users.
- **Solution:** A network of hydrogen fueling stations was established, equipped with polymer electrolyte membrane (PEM) electrolyzers. These PEM electrolyzers use membranes to separate hydrogen from water through electrolysis, providing on-site hydrogen production.
- **Outcome:** Increased accessibility for hydrogen fuel cell vehicle users, reduced dependence on centralized hydrogen production, and support for the growth of the hydrogen transportation sector.

Case Study 2: On-site Hydrogen Generation for Fleets

- **Industry:** Logistics and Transportation.
- **Challenge:** Fleet operators looking to transition to hydrogen-powered vehicles required a reliable and cost-effective solution for on-site hydrogen generation.
- **Solution:** Hydrogen fueling stations were set up at fleet depots, utilizing proton exchange membrane (PEM) electrolyzers. These systems employ membranes to facilitate the electrolysis process, producing hydrogen for the fleet's vehicles.
- **Outcome:** Reduced logistical challenges, lower operating costs, and increased sustainability for fleet operations through on-site hydrogen production.

Case Study 3: Hydrogen Fueling Station Resilience

- **Location:** Japan.
- **Challenge:** Ensuring the resilience of hydrogen fueling stations in the face of natural disasters, such as earthquakes, to maintain continuous service.
- **Solution:** Hydrogen fueling stations in earthquake-prone regions were equipped with metal hydride storage systems. These systems use membranes to regulate the release of hydrogen for fueling, providing a safe and reliable storage solution.
- **Outcome:** Enhanced resilience, uninterrupted service during and after seismic events, and increased confidence in hydrogen fuel cell vehicle adoption.

Practical Example: Multi-Modal Hydrogen Fueling Hub

- **Location:** Europe.
- **Scenario:** A hydrogen fueling hub serves multiple transportation modes, including buses, trucks, and passenger vehicles.
- **Technology:** A combination of on-site hydrogen production using PEM electrolyzers and advanced compression and storage systems with polymer-based membranes. These membranes aid in separating and purifying hydrogen for fueling.

- **Result:** A versatile fueling hub accommodating various hydrogen-powered vehicles, promoting the integration of hydrogen across different transportation sectors.

Emerging Technology: Advanced Hydrogen Dispensing Technologies

- **Innovation:** Research and development focus on advanced hydrogen dispensing technologies, including smart sensors and automated safety systems, to enhance the efficiency and safety of fueling processes.
- **Potential Benefits:** These innovations aim to improve user experience, reduce fueling times, and ensure the safe and reliable dispensing of hydrogen for fuel cell vehicles.

In these examples, the specific membrane technologies employed showcase the role of membranes in on-site hydrogen production, storage, and fueling processes at hydrogen fueling stations. These technologies contribute to the accessibility, sustainability, and safety of hydrogen as a fuel for transportation.

4.1.7 Hydrogen Production

Hydrogen membranes are pivotal in various methods of hydrogen production, including steam methane reforming (SMR) and water electrolysis. The membranes aid in separating and purifying hydrogen, enhancing the overall efficiency and sustainability of hydrogen production processes.

Hydrogen production is a fundamental application where membrane technologies play a crucial role (Dawood et al., 2020). Here are real-world case studies, practical examples, and specific membrane technologies employed for optimal results in hydrogen production:

Case Study 1: Steam Methane Reforming (SMR) Upgrade

- **Industry:** Petrochemical Refining.
- **Challenge:** An existing SMR plant sought to enhance its hydrogen production efficiency and reduce carbon emissions.
- **Solution:** Advanced polymeric membranes with improved hydrogen permeability were integrated into the reforming process. These membranes facilitated the separation of hydrogen from the reformed gas stream, leading to higher purity and increased hydrogen yield.
- **Outcome:** Improved overall efficiency, increased hydrogen production, and reduced environmental impact by minimizing the release of impurities and carbon dioxide.

Case Study 2: Green Hydrogen Production from Renewable Energy

- **Industry:** Renewable Energy.
- **Challenge:** A renewable energy project aimed to produce green hydrogen through water electrolysis using excess energy from wind and solar sources.

- **Solution:** Proton exchange membrane (PEM) electrolyzers were employed, utilizing membranes to facilitate the electrolysis process. These membranes allowed for the selective movement of protons, ensuring the production of high-purity hydrogen.
- **Outcome:** Increased production of green hydrogen, utilizing excess renewable energy, and contributing to the growth of sustainable hydrogen production.

Case Study 3: Biogas Upgrading for Hydrogen Generation

- **Industry:** Waste Management and Renewable Energy.
- **Challenge:** A biogas plant sought to extract hydrogen from biogas for energy generation while upgrading the gas for cleaner utilization.
- **Solution:** Polymeric membranes were implemented to separate and purify hydrogen from the biogas stream. These membranes selectively allowed hydrogen to pass through, leaving behind impurities.
- **Outcome:** Simultaneous biogas upgrading and hydrogen generation, providing a dual benefit of cleaner biogas utilization and hydrogen production for various applications.

Practical Example: On-site Hydrogen Production for Industry

- **Industry:** Various Industrial Processes.
- **Scenario:** Industries requiring a continuous and reliable supply of hydrogen implement on-site production systems to meet their specific needs.
- **Technology:** A combination of PEM electrolyzers and polymer-based gas separation membranes for on-site hydrogen production. These membranes ensure the purification of hydrogen produced through electrolysis.
- **Result:** Enhanced control over hydrogen supply, increased efficiency in industrial processes, and reduced reliance on external hydrogen sources.

Emerging Technology: High-Temperature Electrolysis

- **Innovation:** Ongoing research focuses on high-temperature electrolysis using solid oxide electrolysis cells (SOECs) for hydrogen production.
- **Potential Benefits:** SOECs, with specific ceramic membranes, aim to provide higher efficiency and lower energy consumption in the electrolysis process, contributing to cost-effective and sustainable hydrogen production.

In these examples, the specific membrane technologies employed showcase the versatility of membranes in various hydrogen production methods. These technologies contribute to increased efficiency, sustainability, and adaptability in meeting the diverse demands of hydrogen in different industries.

4.1.8 Hydrogen Recovery from Biogas

In the field of renewable energy, hydrogen membranes are utilized for extracting hydrogen from biogas generated through anaerobic digestion processes. This contributes to cleaner energy production and provides a means of utilizing organic waste for hydrogen generation (Arslan and Yilmaz, 2023).

Hydrogen recovery from biogas is a crucial application that aligns with sustainable and renewable energy goals. Here are real-world case studies, practical examples, and specific membrane technologies employed for optimal results in hydrogen recovery from biogas:

Case Study 1: Municipal Wastewater Treatment Plant

- **Location:** Scandinavia.
- **Challenge:** A municipal wastewater treatment plant sought to enhance its biogas utilization and reduce its environmental footprint.
- **Solution:** Polymeric membranes were implemented to recover hydrogen from the biogas produced during anaerobic digestion. These membranes selectively allowed hydrogen to pass through, leaving behind methane and other impurities.
- **Outcome:** Increased energy output from biogas, improved environmental sustainability, and the production of a high-purity hydrogen stream for various applications.

Case Study 2: Agricultural Biogas Facility

- **Location:** North America
- **Challenge:** An agricultural biogas facility aimed to optimize its biogas upgrading process while recovering valuable hydrogen for additional energy generation.
- **Solution:** Metal hydride storage systems with membranes for controlled hydrogen release were integrated into the biogas upgrading process. These membranes regulated the release of hydrogen, providing a safe and efficient storage solution.
- **Outcome:** Simultaneous biogas upgrading and hydrogen recovery, increased energy yield, and reduced reliance on external energy sources.

Case Study 3: Landfill Gas-to-Energy Project

- **Location:** Australia.
- **Challenge:** A landfill gas-to-energy project sought to maximize the energy potential of the gases emitted from decomposing waste.
- **Solution:** Ceramic membranes were employed for hydrogen separation and purification from the landfill gas stream. These membranes enhanced the quality of the recovered hydrogen for utilization in fuel cells or other energy applications.

- **Outcome:** Increased energy generation from landfill gas, reduced greenhouse gas emissions, and efficient recovery of high-purity hydrogen.

Practical Example: Biogas Upgrading Plant

- **Industry:** Renewable Energy and Waste Management.
- **Scenario:** A biogas upgrading plant processes organic waste to produce biogas for energy use and recovers hydrogen for additional applications.
- **Technology:** Composite membranes combining polymeric and metallic components for efficient hydrogen separation and purification. These membranes ensure a high-purity hydrogen product for various downstream applications.
- **Result:** Improved waste-to-energy conversion, increased energy output, and a sustainable solution for managing organic waste while producing high-quality hydrogen.

Emerging Technology: Bioelectrochemical Systems

- **Innovation:** Research is ongoing in the development of bio electrochemical systems for enhanced hydrogen recovery from biogas. These systems utilize biological processes and specific membranes to selectively extract hydrogen from complex gas mixtures.
- **Potential Benefits:** Bio electrochemical systems aim to provide an energy-efficient and environmentally friendly approach to hydrogen recovery, contributing to the circular economy of waste-to-energy processes.

In these examples, the specific membrane technologies employed showcase the diverse approaches to recovering hydrogen from biogas. These technologies contribute to the efficient utilization of organic waste for energy production and highlight the potential of hydrogen as a clean and renewable energy carrier.

4.1.9 Hydrogen in Energy Storage

Hydrogen membranes play a role in energy storage applications, where excess renewable energy is used to produce hydrogen through electrolysis. The stored hydrogen can later be used for power generation or as a feedstock for various industrial processes, contributing to grid stability and energy resilience.

Hydrogen's role in energy storage is critical for balancing intermittent renewable energy sources and ensuring a stable power supply. Here are real-world case studies, practical examples, and specific membrane technologies employed for optimal results in utilizing hydrogen for energy storage:

Case Study 1: Renewable Energy Microgrid

- **Location:** Europe.
- **Challenge:** A community microgrid integrating renewable energy sources faced the challenge of efficiently storing excess energy for periods of low generation.

- **Solution:** Proton exchange membrane (PEM) electrolyzers were employed to convert excess renewable energy into hydrogen during peak production times. The produced hydrogen was stored in metal hydride storage systems with membranes to regulate hydrogen release.
- **Outcome:** Enhanced energy storage capacity, grid stability, and the ability to balance energy supply and demand within the microgrid.

Case Study 2: Industrial Renewable Hydrogen Storage

- **Industry:** Manufacturing
- **Challenge:** A large manufacturing facility with intermittent renewable energy availability needed a reliable energy storage solution to support its operations during energy lulls.
- **Solution:** Hydrogen produced through high-temperature electrolysis using solid oxide electrolysis cells (SOECs) was stored in advanced porous materials like MOFs equipped with selective membranes. These membranes aided in the efficient and reversible storage of hydrogen.
- **Outcome:** Improved energy resilience, reduced reliance on conventional power sources, and increased sustainability in industrial manufacturing processes.

Case Study 3: Off-Grid Hydrogen-Powered Community

- **Location:** Remote Community
- **Challenge:** A remote community without access to a reliable power grid sought an off-grid energy solution utilizing locally available renewable resources.
- **Solution:** Solar and wind power systems were used to generate hydrogen through electrolysis. The produced hydrogen was stored in advanced composite membranes to maintain purity. Fuel cells were then used to convert stored hydrogen back into electricity as needed.
- **Outcome:** Establishment of a self-sufficient, off-grid community, reduced dependence on external power sources, and increased energy reliability.

Practical Example: Hydrogen Storage for Uninterrupted Power Supply

- **Industry:** Data Center Operations
- **Scenario:** A data center, requiring a continuous and uninterrupted power supply, sought an energy storage solution that could seamlessly integrate with its existing infrastructure.
- **Technology:** Hydrogen produced through PEM electrolysis during low-demand periods was stored in metal hydride storage systems. Membranes were employed to regulate hydrogen release for fuel cells, ensuring a stable and reliable power supply during peak demand.
- **Result:** Uninterrupted data center operations, improved energy efficiency, and reduced reliance on traditional backup power sources.

Emerging Technology: Liquid Organic Hydrogen Carriers (LOHC)

- **Innovation:** Research is ongoing in the development of LOHC systems for hydrogen storage, utilizing carrier liquids to chemically bind and release hydrogen.
- **Potential Benefits:** LOHC systems with specific membranes aim to provide a safe, energy-dense, and reversible hydrogen storage solution, offering advantages in terms of transport, distribution, and energy system integration.

In these examples, the specific membrane technologies employed demonstrate the diverse ways in which hydrogen is utilized for energy storage, showcasing its potential to revolutionize how we store and utilize renewable energy. Membranes play a crucial role in ensuring the efficiency, safety, and reliability of these hydrogen energy storage systems.

4.1.10 Hydrogen in Industrial Processes

Hydrogen membranes find applications in numerous industrial processes, such as hydrogenation in the food industry and reducing atmospheres in metal production. Their ability to provide a controlled and pure hydrogen environment enhances these processes' efficiency and environmental sustainability.

The integration of hydrogen in industrial processes is a versatile application with numerous benefits, including enhanced efficiency, reduced emissions, and increased sustainability. Here are real-world case studies, practical examples, and specific membrane technologies employed for optimal results in utilizing hydrogen in industrial processes:

Case Study 1: Steel Manufacturing

- **Industry:** Steel Production.
- **Challenge:** Traditional steel manufacturing processes involve the use of carbon-intensive methods, contributing to greenhouse gas emissions.
- **Solution:** Hydrogen is employed as a reducing agent in the direct reduction of iron ore. Ceramic membranes are utilized to separate and purify hydrogen, ensuring a clean and efficient process.
- **Outcome:** Reduced carbon emissions, increased energy efficiency, and the production of low-carbon or green steel.

Case Study 2: Ammonia Synthesis

- **Industry:** Chemical Manufacturing (Ammonia Production).
- **Challenge:** Conventional ammonia synthesis relies on hydrogen derived from fossil fuels, contributing to carbon emissions.
- **Solution:** Hydrogen produced through renewable methods, such as electrolysis, is used in ammonia synthesis. Polymer-based membranes contribute to the purification of hydrogen for high-quality ammonia production.

- **Outcome:** Sustainable ammonia production reduced environmental impact, and a shift toward green and renewable feedstocks.

Case Study 3: Hydrogenation in Food Processing

- **Industry:** Food Manufacturing.
- **Challenge:** Hydrogenation processes in food manufacturing often involve the use of hydrogen from fossil sources.
- **Solution:** Utilization of hydrogen produced through electrolysis with PEM electrolyzers. Polymer-based membranes assist in the purification of hydrogen used in the hydrogenation of oils, leading to a more sustainable food processing method.
- **Outcome:** Reduced carbon footprint in food processing, increased use of renewable hydrogen, and improved sustainability in the food industry.

Practical Example: Chemical Synthesis in Pharmaceuticals

- **Industry:** Pharmaceutical Manufacturing.
- **Scenario:** Pharmaceutical companies require high-purity hydrogen for certain chemical synthesis processes.
- **Technology:** A combination of advanced metal membranes and purification technologies is employed to ensure the quality of hydrogen used in pharmaceutical manufacturing. These membranes selectively separate and purify hydrogen for precise chemical reactions.
- **Result:** Improved product quality, compliance with regulatory standards, and a more sustainable approach to pharmaceutical production.

Emerging Technology: Membrane Reactors for Hydrogenation

- **Innovation:** Ongoing research focuses on the development of membrane reactors that integrate catalytic reactions and hydrogen separation within a single unit.
- **Potential Benefits:** Membrane reactors have the potential to enhance the efficiency of hydrogen-involved industrial processes by continuously removing hydrogen from the reaction zone, driving the reaction toward completion, and increasing overall process efficiency.

In these examples, the specific membrane technologies employed showcase the adaptability of hydrogen in various industrial applications. These technologies contribute to increased sustainability, reduced environmental impact, and improved efficiency in industrial processes by integrating renewable hydrogen sources and advanced membrane technologies.

4.2 CHALLENGES IN HYDROGEN MEMBRANE TECHNOLOGY

Hydrogen membrane technology, while promising for various applications, faces several challenges that need to be addressed for widespread adoption and optimal performance (Ball and Wietschel, 2009). Here are the key challenges in hydrogen membrane technology:

4.2.1 Material Selection

- **Challenge:** The challenge of material selection in hydrogen membrane technology revolves around identifying and developing materials with specific properties that are crucial for effective hydrogen separation. These properties include high hydrogen permeability, selectivity, durability, and resistance to degradation under operating conditions.
- **Impact:** The impact of inadequate material selection can be significant. If the chosen materials lack the necessary properties, it can result in membranes with low efficiency, reduced lifespan, and increased maintenance requirements. The success of hydrogen membrane technology relies heavily on the careful selection or design of materials to ensure optimal performance and durability.

4.2.1.1 Considerations and Solutions

Permeability and Selectivity:

- **Polymeric Membranes:** Polymers like polyimides and polybenzimidazoles are commonly used for their high permeability to hydrogen.
- **Ceramic Membranes:** Materials like palladium and its alloys exhibit high selectivity for hydrogen due to their unique permeation properties.

Durability:

- **High Temperatures:** For applications involving high temperatures, materials like ceramics and certain metals (alloys) are preferred to ensure structural integrity.
- **Mechanical Stress:** Membranes must withstand mechanical stress without losing their permeation properties. Ceramic and metal membranes are often chosen for their mechanical robustness.

Resistance to Degradation:

- **Hydrogen Embrittlement:** Certain materials, like metals, are prone to hydrogen embrittlement, a phenomenon where exposure to hydrogen can lead to reduced ductility and mechanical strength. Researchers work on developing alloys that resist this embrittlement.
- **Corrosion:** Corrosion-resistant materials, including specific alloys and coatings, are employed to combat degradation caused by exposure to hydrogen and other gases.

Compatibility:

- **Chemical Compatibility:** Materials must be compatible with the specific gases present in the application. This consideration is crucial to prevent chemical reactions that could compromise membrane performance.

- **Gaseous Impurities:** Some applications involve gases other than hydrogen. The chosen materials should not be adversely affected by the presence of impurities in the gas stream.

Solutions:

Advanced Composite Materials:

- **Polymer-Metal Composites:** Combining the advantages of both polymers and metals to create hybrid materials that exhibit improved selectivity and durability.
- **Ceramic-Polymeric Composites:** Integrating the benefits of ceramics and polymers to achieve a balance between permeability and mechanical strength.

Innovative Nanostructured Materials:

- **Nanostructured Membranes:** Materials with nanoscale structures are explored to enhance permeability and selectivity. For example, carbon nanotubes, graphene, and nanoporous materials show promise in improving membrane performance.

Metal-Organic Frameworks (MOFs):

- MOFs are a class of materials with high porosity and tunable properties. Research focuses on integrating MOFs into membranes to enhance their gas separation capabilities.

Ongoing Research in Material Science:

- Researchers continuously explore new materials and fabrication techniques. This includes experimenting with novel substances, such as ionic liquids and organic-inorganic hybrid materials.

In-Depth Characterization:

- Rigorous characterization techniques, including spectroscopy, microscopy, and mechanical testing, are employed to understand the behavior of materials under hydrogen exposure and other operating conditions.

4.2.2 Membrane Performance

- **Challenge**: Achieving consistent and high-performance levels in various operational conditions, including temperature and pressure variations.
- **Impact:** Inconsistent performance may hinder the reliability and efficiency of hydrogen separation processes.

4.2.2.1 Considerations and Solutions

- **Temperature Sensitivity:**
  Challenge: Membrane performance can be sensitive to temperature changes. Variations in operating temperatures may lead to fluctuations in separation efficiency.
  Solution: Research focuses on developing membranes that maintain stable performance over a wide temperature range. This involves the use of materials with temperature-resistant properties, as well as innovative designs that accommodate thermal variations.
- Pressure Effects:
  Challenge: Pressure fluctuations in the feed gas can impact membrane performance. Sudden changes in pressure may affect selectivity and permeability, leading to reduced separation efficiency.
  Solution: Membrane engineering considers the development of materials and structures that can withstand varying pressures. Additionally, the incorporation of pressure-resistant components helps maintain consistent performance.
- Robustness to Operational Conditions:
  Challenge: Industrial applications often involve dynamic operational conditions. Membranes must perform reliably under varying pressures, temperatures, and gas compositions.
  Solution: Rigorous testing under simulated operational conditions allows researchers to assess membrane performance and identify areas for improvement. Advances in material science contribute to the development of robust membranes that can withstand diverse industrial environments.
- Scalability:
  Challenge: Transitioning from laboratory-scale testing to industrial-scale applications introduces new challenges in maintaining membrane performance at larger scales.
  Solution: Pilot projects and scaled testing facilities provide valuable insights into how membranes perform under real-world conditions. Iterative testing and optimization are essential to ensure scalability without sacrificing efficiency.
- Integration with Other Process Components:
  Challenge: Membranes are often part of larger hydrogen separation systems that include compressors, heat exchangers, and other components. Ensuring seamless integration is crucial for overall system performance.
  Solution: Engineering solutions that optimize the coordination between membrane units and ancillary equipment. This involves designing systems that can adapt to fluctuations in temperature, pressure, and flow rates.
- Dynamic Feed Gas Composition:
  Challenge: Changes in the composition of the feed gas, including impurities or fluctuations in hydrogen concentration, can impact membrane performance.
  Solution: Developing membranes with enhanced impurity tolerance and adapting system controls to respond dynamically to changes in feed gas

composition. This helps maintain consistent separation efficiency despite variations in gas quality.

- In-Service Monitoring and Control:
  Challenge: Ensuring continuous monitoring and control of membrane performance during operation.
  Solution: Implementation of advanced sensing and control systems that allow real-time monitoring of key parameters. This enables immediate adjustments to maintain optimal membrane performance.

Addressing the challenges related to membrane performance requires a holistic approach, considering the material properties, system design, and operational conditions. Through ongoing research and technological advancements, the goal is to enhance the reliability and efficiency of hydrogen separation processes across diverse applications.

4.2.3 Hydrogen Embrittlement

- **Challenge:** Hydrogen can cause embrittlement in certain materials, leading to mechanical failure and reduced structural integrity.
- **Impact:** Embrittlement can compromise the durability and safety of membranes, especially in high-stress environments.

4.2.3.1 Considerations and Solutions

Mechanisms of Hydrogen Embrittlement:

Challenge: Understanding the mechanisms by which hydrogen embrittlement occurs is crucial. Hydrogen can penetrate materials, causing microstructural changes that result in reduced ductility and mechanical strength.
Solution: In-depth research into the fundamental processes of hydrogen embrittlement helps identify susceptible materials and develop strategies to mitigate its effects.

Material Selection:

Challenge: Certain materials, such as metals and alloys, are more prone to hydrogen embrittlement. Choosing materials with inherent resistance to embrittlement is a key consideration.
Solution: Opting for materials that have demonstrated resistance to hydrogen embrittlement, or developing alloys specifically designed to mitigate embrittlement effects.

Preventing Hydrogen Uptake:

Challenge: Preventing or minimizing the uptake of hydrogen by materials is essential to reduce the risk of embrittlement.

Solution: Incorporating protective coatings, modifying material surfaces, or using barriers that limit hydrogen diffusion can be effective strategies to prevent excessive hydrogen absorption.

Operational Parameters:

Challenge: Operating conditions, including temperature and pressure, can influence the severity of hydrogen embrittlement.
Solution: Optimizing operational parameters to minimize the impact of embrittlement. This involves selecting appropriate temperature and pressure ranges that balance performance and safety.

Monitoring and Inspection:

Challenge: Detecting signs of hydrogen embrittlement in membranes can be challenging, especially in situ.
Solution: Implementing monitoring and inspection techniques, such as non-destructive testing methods, to assess the structural integrity of membranes over time. Early detection allows for preventive measures or timely replacements.

Material Modification Techniques:

Challenge: Developing techniques to modify materials and make them less susceptible to hydrogen embrittlement.
Solution: Utilizing material modification methods, such as alloying, heat treatment, or surface engineering, to enhance resistance to hydrogen embrittlement while maintaining other desired properties.

Research into Hydrogen Interaction Mechanisms:

Challenge: A comprehensive understanding of the interaction between hydrogen and membrane materials is crucial for devising effective prevention strategies.
Solution: Ongoing research into the molecular-level interactions between hydrogen and materials, including the use of computational modeling and simulation, helps inform strategies to mitigate embrittlement.

Hydrogen Compatibility Testing:

Challenge: Ensuring that membrane materials are thoroughly tested for hydrogen compatibility before deployment.
Solution: Conducting extensive compatibility testing under conditions representative of the intended application to identify potential issues related to embrittlement.

Addressing hydrogen embrittlement involves a combination of material science advancements, thorough testing protocols, and the development of preventive measures. By minimizing the risks associated with embrittlement, hydrogen membrane technologies can achieve greater durability and safety in diverse operating environments.

4.2.4 Membrane Stability

- **Challenge:** Ensuring long-term stability and reliability of membranes, particularly in applications with continuous and prolonged operation.
- **Impact:** Membrane degradation over time can result in decreased efficiency, increased maintenance, and replacement costs.

4.2.4.1 Considerations and Solutions

Material Durability:

Challenge: The materials used in membranes must withstand prolonged exposure to operational conditions without experiencing degradation.
Solution: Selecting materials with high durability, resistance to chemical or thermal degradation, and long-term stability under the specific environment in which the membrane operates.

Operational Conditions:

Challenge: Membranes are exposed to a variety of operational conditions, including temperature fluctuations, pressure changes, and exposure to potentially corrosive gases.
Solution: Designing membranes that can tolerate a wide range of operational conditions through careful material selection and engineering solutions, ensuring stability in diverse environments.

Chemical Compatibility:

Challenge: The presence of impurities or reactive gases in the feed stream can lead to chemical reactions with membrane materials, causing degradation.
Solution: Choosing materials that are chemically compatible with the specific gas mixtures encountered in the application, or implementing pre-treatment steps to remove reactive components before reaching the membrane.

Membrane Fouling:

Challenge: Over time, membranes may accumulate deposits or contaminants, reducing their efficiency and stability.
Solution: Incorporating anti-fouling measures, such as surface coatings, periodic cleaning protocols, or integrating pre-filtration stages to prevent fouling and maintain long-term stability.

Oxidative Stability:

Challenge: Exposure to oxygen or oxidative gases can lead to membrane degradation, particularly in certain polymeric materials.
Solution: Choosing materials with high oxidative stability or incorporating antioxidant additives to mitigate the impact of oxidative stress on membrane performance.

Research in Advanced Materials:

Challenge: The development of new materials with enhanced stability and performance over extended periods.
Solution: Continuous research in material science to discover and design advanced materials, such as nanostructured composites or self-healing polymers, that exhibit improved stability under challenging conditions.

Life Cycle Assessments:

Challenge: Predicting the long-term performance of membranes requires comprehensive life cycle assessments that consider factors like material degradation, fouling, and maintenance needs.
Solution: Conducting thorough life cycle assessments to evaluate the overall durability and stability of membranes under realistic operating scenarios, and guiding improvements in design and material selection.

Condition Monitoring and Predictive Maintenance:

Challenge: Identifying early signs of membrane degradation before it significantly impacts performance.
Solution: Implementing condition monitoring systems and predictive maintenance protocols, such as real-time sensors and analytics, to detect potential issues and intervene before stability is compromised.

Ensuring membrane stability is a multifaceted challenge that involves a combination of material science advancements, robust engineering designs, and proactive maintenance strategies. By addressing these considerations, membrane technologies can offer long-term reliability and contribute to the sustained efficiency of hydrogen separation processes.

4.2.5 Scaling-Up Problem

Challenge: Successfully scaling up membrane technologies from laboratory-scale to industrial applications while maintaining performance and cost-effectiveness.
Impact: Scaling issues can impede the feasibility of large-scale hydrogen production or separation projects.

4.2.5.1 Considerations and Solutions

Materials and Manufacturing Consistency:

Challenge: Achieving consistency in material properties and manufacturing processes when scaling up from laboratory to industrial-scale production.
Solution: Developing stringent quality control measures and manufacturing processes that ensure uniformity in material composition and membrane fabrication. Continuous monitoring and optimization are crucial during the scaling-up process.

System Integration Challenges:

Challenge: Integrating larger membrane systems into existing or new industrial facilities may pose challenges in terms of system compatibility, space requirements, and overall integration.
Solution: Conducting thorough engineering assessments and system integration studies to ensure seamless incorporation of larger membrane units into industrial processes. This may involve redesigning certain aspects of the overall system.

Economic Feasibility:

Challenge: Achieving cost-effectiveness at an industrial scale is often challenging due to increased material and manufacturing costs.
Solution: Optimizing the manufacturing process, leveraging economies of scale, and exploring innovative cost reduction strategies to make large-scale membrane technologies economically viable.

Energy Efficiency at Scale:

Challenge: Maintaining or improving energy efficiency as the scale of hydrogen production or separation increases.
Solution: Implementing design modifications and process optimizations to ensure that energy consumption remains competitive at larger scales. This may involve enhancements in heat recovery, integration with renewable energy sources, and overall system efficiency improvements.

Mass Transfer Limitations:

Challenge: Scaling up can lead to challenges in mass transfer limitations, affecting the efficiency of hydrogen separation.
Solution: Conducting detailed mass transfer studies during the scaling-up process to identify and address potential limitations. Adjusting membrane design or system configurations to optimize mass transfer at larger scales.

Reliability and Maintenance:

Challenge: Maintaining reliability and ease of maintenance as the number and size of membrane units increase.
Solution: Designing systems with redundancy, implementing predictive maintenance strategies, and ensuring accessibility for routine inspections and repairs to minimize downtime.

Environmental and Safety Considerations:

Challenge: Scaling up may introduce new environmental and safety considerations that need to be addressed to comply with regulations and ensure safe operation.
Solution: Conducting comprehensive environmental impact assessments and safety evaluations during the scaling-up process. Implementing necessary safety measures and obtaining regulatory approvals to meet industry standards.

Pilot Testing and Validation:

Challenge: Scaling up without adequate pilot testing may lead to unforeseen challenges and operational issues.
Solution: Conducting thorough pilot testing at intermediate scales to validate performance, identify potential challenges, and fine-tune the system before full-scale implementation.

Addressing the scaling-up problem involves a combination of engineering expertise, systematic testing, and continuous improvement strategies. Successful scaling requires collaboration between researchers, engineers, and industry stakeholders to navigate the complexities associated with transitioning from laboratory-scale prototypes to large-scale industrial applications.

4.2.6 Impurity Tolerance

- **Challenge:** Developing membranes that can tolerate impurities commonly found in industrial gas streams without degradation or loss of performance.
- **Impact:** Impurity intolerance may require additional purification steps, increasing complexity and costs.

4.2.6.1 Considerations and Solutions

Identification of Common Impurities:

Challenge: Industrial gas streams often contain impurities such as sulfur compounds, carbon dioxide, water vapor, and trace contaminants.
Solution: Comprehensive analysis and identification of common impurities help in designing membranes that can withstand exposure to these specific substances.

Material Selection for Impurity Resistance:

Challenge: Impurities can react with or adsorb onto membrane materials, leading to degradation.
Solution: Choosing materials with inherent resistance to impurities or developing coatings that protect membrane surfaces from impurity-induced degradation.
Materials such as corrosion-resistant alloys or specialized polymer blends may be employed.

Innovative Coating Technologies:

Challenge: Developing coatings that enhance impurity resistance without compromising membrane permeability.
Solution: Research and development of advanced coatings, such as thin films or surface modifications, to create impurity-resistant barriers while maintaining or improving membrane performance.

Hybrid Membrane Systems:

Challenge: Achieving impurity tolerance without sacrificing overall system efficiency.
Solution: Integrating hybrid membrane systems that combine different membrane types, each specialized in handling specific impurities. This allows for a comprehensive approach to impurity tolerance without compromising the overall performance of the system.

Pre-treatment Technologies:

Challenge: Impurity-rich gas streams may require pre-treatment to remove harmful substances before reaching the membrane.
Solution: Implementing pre-treatment technologies, such as adsorption beds or chemical scrubbers, to remove impurities upstream of the membrane. This helps in protecting the membrane and maintaining long-term stability.

Selective Permeation Enhancement:

Challenge: Enhancing the selectivity of membranes to selectively allow the passage of hydrogen while blocking impurities.
Solution: Designing membranes with enhanced selectivity through material engineering or modifications to preferentially allow hydrogen permeation while restricting the passage of impurities.

Real-time Monitoring Systems:

Challenge: Ensuring timely detection of impurity-induced degradation to enable proactive maintenance.
Solution: Implementing real-time monitoring systems that can detect changes in membrane performance or integrity. This allows for prompt intervention or adjustment to minimize the impact of impurities on membrane function.

Process Optimization:

Challenge: Balancing impurity tolerance with overall system efficiency and cost-effectiveness.
Solution: Continuously optimizing the hydrogen separation process through system adjustments, operational parameter tuning, and material enhancements to achieve an optimal balance between impurity tolerance and performance.

Regulatory Compliance:

Challenge: Meeting regulatory standards for the permissible levels of impurities in hydrogen products.
Solution: Collaborating with regulatory bodies and incorporating necessary monitoring and control measures to ensure compliance with established standards for impurity levels.
Successfully addressing the challenge of impurity tolerance involves a combination of material science innovations, engineering strategies, and system-level optimizations to create robust and efficient hydrogen membrane technologies for industrial applications.

4.2.7 Temperature and Pressure Management

- **Challenge:** Managing and optimizing membrane performance under varying temperature and pressure conditions.
- **Impact:** Inadequate temperature and pressure management can affect separation efficiency and overall system reliability.

4.2.7.1 Considerations and Solutions

Temperature Sensitivity:

- **Challenge:** Membrane performance is often sensitive to temperature changes, and variations can impact separation efficiency.
- **Solution:** Implementing temperature control mechanisms, such as insulation or integrated heating/cooling systems, to maintain optimal membrane operating temperatures. Additionally, developing membranes that exhibit stable performance across a range of temperatures.

Pressure Fluctuations:

- **Challenge:** Sudden changes in pressure can affect membrane selectivity and permeability, influencing separation efficiency.
- **Solution:** Incorporating pressure control systems and employing membranes designed to withstand varying pressure conditions. Ensuring the structural integrity of the membrane under different pressure levels is essential for consistent performance.

Thermal Cycling Considerations:

- **Challenge:** Industrial processes may involve thermal cycling, exposing membranes to repeated temperature changes.
- **Solution:** Designing membranes and systems that can withstand thermal cycling without compromising performance. This may involve the use of materials with low thermal expansion coefficients and robust structural designs.

Optimal Operating Conditions:

- **Challenge:** Identifying and maintaining the optimal operating conditions for membrane performance.
- **Solution:** Conducting thorough experimental studies to determine the ideal temperature and pressure ranges for specific membrane materials and configurations. Utilizing advanced modeling techniques to predict and optimize performance under varying conditions.

Hybrid Membrane Systems:

- **Challenge:** Achieving optimal performance under a wide range of temperature and pressure conditions may require combining different types of membranes.
- **Solution:** Integrating hybrid membrane systems that leverage the strengths of different membrane types to enhance overall performance across varying operating conditions.

Adaptive Control Systems:

- **Challenge:** Membrane systems need to adapt to fluctuations in temperature and pressure to maintain efficiency.
- **Solution:** Implementing adaptive control systems that continuously monitor operating conditions and adjust parameters in real time. This ensures that the membrane operates within its optimal performance range despite variations in temperature and pressure.

Materials for Extreme Conditions:

- **Challenge:** Some industrial applications involve extreme temperatures and pressures beyond the typical range.
- **Solution:** Developing specialized membranes using materials engineered to withstand extreme conditions. This may involve high-temperature ceramics, advanced alloys, or hybrid materials that offer enhanced durability and stability.

In Situ Monitoring and Sensors:

- **Challenge:** Ensuring real-time monitoring of temperature and pressure conditions at the membrane interface.
- **Solution:** Implementing in situ sensors that provide continuous feedback on environmental conditions. This allows for prompt adjustments to maintain optimal membrane performance.

Energy-Efficient Heating/Cooling Systems:

- **Challenge:** Maintaining temperature control without excessive energy consumption.
- **Solution:** Incorporating energy-efficient heating and cooling systems, such as heat exchangers or thermal management technologies, to optimize temperature control without significantly impacting the overall energy balance of the system.

By addressing the challenges associated with temperature and pressure management, hydrogen membrane technologies can achieve consistent and reliable performance, contributing to their successful implementation in various industrial applications.

4.2.8 Cost and Commercial Viability

- **Challenge:** Achieving cost-effective production, installation, and maintenance of hydrogen membrane systems for commercial viability.
- **Impact:** High costs can hinder widespread adoption and competitiveness with other hydrogen production or separation methods.

4.2.8.1 Considerations and Solutions

Material Costs:

- **Challenge:** The cost of materials used in membrane fabrication can significantly impact overall system costs.
- **Solution:** Exploring cost-effective materials without compromising performance. Additionally, researching alternative materials or fabrication techniques that reduce material expenses.

Manufacturing Processes:

- **Challenge:** The complexity of manufacturing processes can contribute to higher production costs.
- **Solution:** Streamlining manufacturing processes, adopting automation where feasible, and optimizing production workflows to reduce labor and energy costs.

Economies of Scale:

- **Challenge:** Small-scale production may result in higher per-unit costs.
- **Solution:** Scaling up production to benefit from economies of scale, which can reduce costs per unit by spreading fixed expenses over a larger production volume.

Research and Development Investment:

- **Challenge:** Significant research and development investments are often required to advance membrane technologies.
- **Solution:** Collaborating with research institutions, industry partners, and government initiatives to share R&D costs. Leveraging public-private partnerships can help accelerate technology development and reduce financial burdens.

System Integration Costs:

- **Challenge:** Integrating membrane systems into existing industrial processes may incur additional costs.
- **Solution:** Conducting thorough feasibility studies and system integration assessments to identify cost-effective integration strategies. Collaborating with industry partners to develop standardized interfaces that facilitate seamless integration.

Long-Term Maintenance and Operational Costs:

- **Challenge:** High maintenance costs can impact the long-term economic viability of membrane systems.
- **Solution:** Designing membranes and systems for durability and ease of maintenance. Implementing predictive maintenance strategies and utilizing materials with longer lifespans can help reduce operational costs over time.

Energy Efficiency:

- **Challenge:** Energy consumption contributes significantly to operational costs.
- **Solution:** Continuously optimizing membrane systems for energy efficiency. Integrating energy recovery systems, utilizing renewable energy sources, and implementing advanced control strategies can help reduce energy consumption and costs.

Market Competition and Pricing:

- **Challenge:** Competing with established hydrogen production and separation methods with lower upfront costs.
- **Solution:** Offering unique selling points, such as improved efficiency, lower environmental impact, or specific application advantages, to justify the initial investment. Collaborating with industry partners to explore competitive pricing models.

Regulatory and Incentive Programs:

- **Challenge:** Navigating regulatory requirements and compliance standards can add complexity and costs.
- **Solution:** Engaging with regulatory bodies to streamline approval processes and participating in incentive programs or subsidies that promote the adoption of clean hydrogen technologies.

Life Cycle Cost Analysis:

- **Challenge:** Focusing on upfront costs without considering the entire life cycle can lead to suboptimal decisions.
- **Solution:** Conducting comprehensive life cycle cost analyses that account for production, installation, maintenance, and decommissioning costs. This holistic approach provides a more accurate assessment of commercial viability.

By addressing these considerations, hydrogen membrane technologies can strive to achieve cost-effectiveness, making them more competitive and attractive for widespread adoption in various industrial applications.

4.2.9 Integration of Membranes into Practical Separation Systems

- **Challenge:** Developing efficient and practical integration strategies for membrane technologies within existing or new hydrogen separation systems.
- **Impact:** Poor integration may result in suboptimal performance and increased complexity in industrial applications.

4.2.9.1 Considerations and Solutions

System Compatibility:

- **Challenge:** Ensuring that membrane systems are compatible with existing or new industrial processes.
- **Solution:** Conducting comprehensive compatibility assessments and engineering studies to identify potential integration challenges. Modifying or adapting membrane systems to seamlessly fit within the existing process infrastructure.

Standardization of Interfaces:

- **Challenge:** Lack of standardized interfaces may lead to difficulties in integrating membrane systems with other components.
- **Solution:** Collaborating with industry stakeholders to establish standard interfaces for membrane systems. This facilitates easier integration, reduces engineering costs, and promotes interoperability.

Modular Design:

- **Challenge:** Traditional systems may not easily accommodate modular membrane units.
- **Solution:** Designing membranes and associated components in a modular fashion, allowing for easy integration and scalability. Modular systems enable the addition or removal of membrane units based on capacity requirements.

Optimizing Flow Dynamics:

- **Challenge:** Achieving optimal flow dynamics within the integrated system to maximize separation efficiency.
- **Solution:** Conducting computational fluid dynamics (CFD) simulations and experimental studies to optimize the layout and configuration of membranes within the system. Implementing well-designed flow patterns to enhance mass transfer and minimize pressure drop.

Parallel and Series Configurations:

- **Challenge:** Determining the most effective configuration for membrane units within the system.
- **Solution:** Evaluating the benefits of parallel and series configurations based on the specific application and requirements. Parallel configurations enhance capacity, while series configurations improve selectivity. Designing systems that can adapt to changing operating conditions.

Process Control and Automation:

- **Challenge:** Integrating membrane systems into automated control processes to optimize performance.
- **Solution:** Implementing advanced process control and automation systems that continuously monitor and adjust operational parameters. This ensures real-time optimization of membrane performance and system efficiency.

Waste and Byproduct Handling:

- **Challenge:** Addressing the handling of waste streams or byproducts generated during membrane separation.

- **Solution:** Incorporating efficient waste management strategies, such as recycling or treating byproducts within the overall process. Minimizing environmental impact and ensuring regulatory compliance.

Energy Recovery:

- **Challenge:** Energy efficiency may be compromised without effective integration of energy recovery mechanisms.
- **Solution:** Integrating energy recovery systems, such as pressure exchangers or heat exchangers, to capture and reuse energy within the system. Optimizing the overall energy balance to enhance the economic viability of the membrane technology.

Sensor and Monitoring Systems:

- **Challenge:** Lack of real-time data may hinder efficient operation and maintenance.
- **Solution:** Implementing advanced sensor and monitoring systems that provide real-time data on membrane performance, pressure, temperature, and other key parameters. This allows for proactive decision-making and maintenance planning.

Pilot Testing and Iterative Optimization:

- **Challenge:** Implementing a new membrane system without thorough testing may lead to unforeseen challenges.
- **Solution:** Conducting pilot testing at intermediate scales to identify potential integration issues, gather performance data, and iteratively optimize the system design before full-scale implementation.

Effective integration of membrane technologies into hydrogen separation systems requires a multidisciplinary approach, involving collaboration between membrane engineers, process engineers, and system integrators. Continuous optimization and adaptation based on real-world performance data contribute to the successful implementation of membrane systems within practical separation applications.

4.2.10 Regulatory and Safety Considerations

- **Challenge:** Adhering to regulatory standards and ensuring the safety of hydrogen membrane systems in various applications.
- **Impact:** Failure to meet safety and regulatory requirements can impede deployment and public acceptance of hydrogen membrane technologies.

4.2.10.1 Considerations and Solutions

Regulatory Compliance:

- **Challenge:** Navigating and adhering to complex and evolving regulatory frameworks for hydrogen technologies.

- **Solution:** Establishing a dedicated regulatory affairs team to stay informed about current and upcoming regulations. Engaging with regulatory authorities to seek guidance, participate in industry discussions, and ensure compliance with safety and environmental standards.

Safety Standards and Certification:

- **Challenge:** Meeting safety standards and obtaining relevant certifications for membrane systems.
- **Solution:** Collaborating with safety certification bodies to undergo rigorous testing and certification processes. Ensuring that membrane systems comply with industry-specific safety standards and are equipped with safety features to mitigate potential risks.

Risk Assessments:

- **Challenge:** Identifying and mitigating potential risks associated with hydrogen membrane systems.
- **Solution:** Conducting comprehensive risk assessments that evaluate potential hazards, assess the probability and severity of incidents, and implement risk mitigation measures. Regularly updating risk assessments as technologies and applications evolve.

Emergency Response Planning:

- **Challenge:** Developing effective emergency response plans to address potential accidents or system failures.
- **Solution:** Collaborating with local emergency services, developing detailed emergency response plans, and conducting regular drills to ensure that personnel are well-prepared to handle potential incidents. Providing clear guidance on emergency shutdown procedures.

Public Awareness and Education:

- **Challenge:** Lack of public awareness and understanding of hydrogen membrane technologies may lead to concerns.
- **Solution:** Implementing public awareness campaigns to educate communities, local authorities, and stakeholders about the safety measures in place. Sharing information on the benefits and safety features of hydrogen membrane systems to build public confidence.

Material Safety and Compatibility:

- **Challenge:** Ensuring the safety and compatibility of materials used in membrane systems with hydrogen and other gases.

- **Solution:** Conducting material compatibility studies and testing to ensure that materials are resistant to hydrogen embrittlement, corrosion and other potential issues. Choosing materials with a proven track record of safety in hydrogen service.

Regulatory Engagement:

- **Challenge:** Limited collaboration with regulatory authorities may lead to uncertainties and delays.
- **Solution:** Proactively engaging with regulatory agencies, participating in industry forums, and providing input on the development of standards and regulations related to hydrogen membrane technologies. Building constructive relationships with regulatory authorities to facilitate smooth approvals.

Documentation and Record-Keeping:

- **Challenge:** Incomplete or inadequate documentation may hinder regulatory approvals.
- **Solution:** Maintaining meticulous records of design specifications, testing procedures, safety features, and compliance documentation. Ensuring that all relevant information is readily available for regulatory submissions and audits.

Continuous Monitoring and Reporting:

- **Challenge:** Demonstrating ongoing compliance with safety and regulatory standards.
- **Solution:** Implementing continuous monitoring systems to track key parameters and performance indicators. Establishing reporting mechanisms to promptly inform regulatory authorities of any deviations or incidents, demonstrating a commitment to transparency and compliance.

Global Regulatory Variations:

- **Challenge:** Navigating variations in regulatory requirements across different regions and countries.
- **Solution:** Tailoring compliance strategies to meet specific regional regulatory requirements. Collaborating with local authorities and regulatory bodies in each jurisdiction to ensure alignment with safety and environmental standards.

By prioritizing safety considerations and actively engaging with regulatory bodies, hydrogen membrane technology developers can foster a regulatory environment that supports the safe deployment of these technologies in various applications. This approach enhances public confidence and paves the way for the broader adoption of hydrogen membrane systems.

4.3 PERFORMANCE EVALUATION OF HYDROGEN MEMBRANES

4.3.1 Strategies for Enhancing Selectivity, Permeability, and Stability of Hydrogen Separation Membranes

Hydrogen separation membranes play a crucial role in various industrial applications, and their performance is determined by factors such as selectivity, permeability, and stability. Implementing strategies to enhance these key attributes is essential for optimizing the overall efficiency and reliability of hydrogen membrane systems (Kume et al., 2013). Here are strategies for improving selectivity, permeability, and stability:

1. **Material Innovation:**
   - **Selectivity Enhancement:**
     - *Strategy:* Develop and explore novel membrane materials with inherent selectivity for hydrogen. This may involve advanced polymers, mixed matrix materials, or nanocomposite structures that selectively allow hydrogen to permeate while blocking other gases.
   - **Permeability Improvement:**
     - *Strategy:* Investigate materials with high hydrogen permeability. Tailor material structures to facilitate faster hydrogen transport without compromising selectivity. Utilize advanced synthesis techniques to achieve precise control over material properties.
   - **Stability Enhancement:**
     - *Strategy:* Identify materials with superior stability under various operating conditions. Consider materials resistant to chemical degradation, thermal stress, and embrittlement. Incorporate additives or surface modifications to enhance the overall stability of the membrane.
2. **Membrane Design and Structure:**
   - **Selectivity Enhancement:**
     - *Strategy:* Optimize membrane structure to maximize selectivity. Utilize advanced membrane design principles, such as asymmetric structures or layered configurations, to enhance the preferential permeation of hydrogen over other gases.
   - **Permeability Improvement:**
     - *Strategy:* Fine-tune membrane thickness and morphology to promote efficient hydrogen permeation. Explore membrane architectures, such as thin-film or hollow-fiber configurations, to achieve high permeability while maintaining structural integrity.
   - **Stability Enhancement:**
     - *Strategy:* Design membranes with inherent resistance to degradation. Incorporate structural features that minimize the impact of stressors, such as reinforcing layers or support structures. Explore innovative membrane geometries to enhance stability during operation.

3. **Surface Modifications and Coatings:**
    - **Selectivity Enhancement**:
        - *Strategy:* Apply selective coatings or surface modifications to enhance the preferential adsorption or diffusion of hydrogen. Functionalize membrane surfaces to selectively interact with hydrogen molecules.
    - **Permeability Improvement:**
        - *Strategy:* Employ coatings that reduce surface resistance and enhance hydrogen permeability. Explore surface modification techniques, such as plasma treatment or chemical doping, to optimize membrane surface properties.
    - **Stability Enhancement:**
        - *Strategy:* Implement protective coatings to shield membrane surfaces from harsh operating conditions. Develop coatings with self-healing properties to mitigate the impact of minor defects and extend membrane lifespan.
4. **Process Optimization:**
    - **Selectivity Enhancement:**
        - *Strategy:* Optimize process parameters, such as temperature and pressure, to favor hydrogen selectivity. Conduct detailed studies on the impact of operating conditions on membrane selectivity.
    - **Permeability Improvement:**
        - *Strategy:* Fine-tune operating parameters to maximize hydrogen permeability. Investigate the influence of temperature, pressure, and gas composition on membrane performance and adjust accordingly.
    - **Stability Enhancement:**
        - *Strategy:* Implement controlled operating conditions to minimize stress on the membrane. Develop adaptive process control systems that respond to changing conditions to maintain stability.
5. **Advanced Manufacturing Techniques:**
    - **Selectivity Enhancement:**
        - *Strategy:* Explore advanced manufacturing methods, such as precision casting or 3D printing, to create intricate membrane structures that enhance selectivity.
    - **Permeability Improvement:**
        - *Strategy:* Utilize advanced fabrication techniques to achieve nanoscale features that enhance permeability. Incorporate microfabrication technologies for precise control over membrane structure.
    - **Stability Enhancement:**
        - *Strategy:* Employ manufacturing techniques that produce uniform and defect-free membranes. Implement quality control measures to ensure consistency in membrane fabrication.
6. **Innovative Hybrid Membrane Systems:**
    - **Selectivity Enhancement:**
        - *Strategy:* Integrate different membrane types to capitalize on their individual selectivity strengths. Develop hybrid systems that combine complementary membranes to achieve enhanced overall selectivity.

- **Permeability Improvement:**
  - *Strategy:* Combine membranes with varying permeability characteristics in a hybrid system. Optimize the arrangement of different membranes to achieve synergistic effects on overall hydrogen permeation.
- **Stability Enhancement:**
  - *Strategy:* Utilize hybrid systems with membranes that offer complementary stability features. Design systems where the stability of one membrane type compensates for potential weaknesses in another.

Implementing these strategies requires a multidisciplinary approach, involving materials science, membrane engineering, and process optimization. Continuous research and development efforts focused on these enhancement strategies will contribute to the advancement of hydrogen separation membrane technologies with improved selectivity, permeability, and stability.

Case Study 1: Hydrogen Separation for Industrial Processes

Objective: Enhancing hydrogen purity in an ammonia production plant
Overview: A leading ammonia production facility sought to improve the purity of hydrogen used in its Haber-Bosch process. The conventional purification methods were energy-intensive and required frequent maintenance.
Implementation: The facility integrated advanced hydrogen separation membranes into their existing hydrogen purification system. The selected membranes were designed for high selectivity, allowing hydrogen to permeate while blocking impurities like nitrogen and methane effectively.

Results:

1. **Increased Hydrogen Purity:** The hydrogen separation membranes significantly improved the purity of hydrogen, meeting the stringent requirements for ammonia production.
2. **Energy Efficiency:** Compared to the traditional purification methods, the membrane system demonstrated lower energy consumption, contributing to overall process efficiency.
3. **Reduced Downtime:** The robustness and stability of the membranes led to extended operational periods between maintenance cycles, reducing downtime and enhancing production continuity.

Conclusion: The successful integration of hydrogen separation membranes not only improved the product quality but also resulted in energy savings and operational reliability, making it a cost-effective and sustainable solution for ammonia production.

Case Study 2: Hydrogen Recovery from Biogas

Objective: Recovering hydrogen from biogas generated at a wastewater treatment plant

Overview: A municipal wastewater treatment plant aimed to harness hydrogen from the biogas produced during the anaerobic digestion process. The challenge was to efficiently recover hydrogen while minimizing the presence of impurities.

Implementation: The plant incorporated specialized hydrogen separation membranes designed for biogas applications. These membranes were chosen for their impurity tolerance and ability to operate in variable conditions.

Results:

1. **Biogas Upgrading:** The hydrogen separation membranes effectively separated hydrogen from the biogas, upgrading it to a high-purity methane product for use as renewable natural gas.
2. **Impurity Tolerance:** The membranes demonstrated resilience against impurities commonly found in biogas, reducing the need for additional purification steps.
3. **Green Hydrogen Production:** Recovered hydrogen was used for on-site energy generation or injected into the local gas grid, contributing to the production of green hydrogen.

Conclusion: The successful implementation of hydrogen separation membranes in the biogas upgrading process showcased the technology's adaptability to diverse gas compositions, offering a sustainable solution for hydrogen recovery and green energy production.

These case studies illustrate the practical applications and benefits of hydrogen membrane technology in industrial settings, showcasing its versatility in improving efficiency, purity, and sustainability across various processes.

REFERENCES

Amin, M., Butt, A.S., Ahmad, J., Lee, C., Azam, S.U., Mannan, H.A., Naveed, A.B., Farooqi, Z.U.R., Chung, E., Iqbal, A., 2023. Issues and challenges in hydrogen separation technologies. *Energy Rep* 9, 894–911.

Arslan, M., Yilmaz, C., 2023. Investigation of green hydrogen production and development of waste heat recovery system in biogas power plant for sustainable energy applications. *Int J Hydrogen Energy* 48, 26652–26664.

Ball, M., Wietschel, M., 2009. The future of hydrogen--opportunities and challenges. *Int J Hydrogen Energy* 34, 615–627.

Barthélémy, H., Weber, M., Barbier, F., 2017. Hydrogen storage: recent improvements and industrial perspectives. *Int J Hydrogen Energy* 42, 7254–7262.

Bernardo, G., Araújo, T., da Silva Lopes, T., Sousa, J., Mendes, A., 2020. Recent advances in membrane technologies for hydrogen purification. *Int J Hydrogen Energy* 45, 7313–7338.

Cardona, P., Costa-Castelló, R., Roda, V., Carroquino, J., Valiño, L., Serra, M., 2023. Model predictive control of an on-site green hydrogen production and refuelling station. *Int J Hydrogen Energy* 48, 17995–18010.

Dawood, F., Anda, M., Shafiullah, G.M., 2020. Hydrogen production for energy: an overview. *Int J Hydrogen Energy* 45, 3847–3869.

Dube, S., Gorimbo, J., Moyo, M., Okoye-Chine, C.G., Liu, X., 2023. Synthesis and application of Ni-based membranes in hydrogen separation and purification systems: a review. *J Environ Chem Eng* 11, 109194.

Faraji, S., Sotudeh-Gharebagh, R., Mostoufi, N., 2005. Hydrogen recovery from refinery off-gases. *J Appl Sci (Pakistan)* 5, 459–464.

Gamboa, A., Fernandes, E.C., 2024. Resistive hydrogen sensors based on carbon nanotubes: a review. *Sens Actuators A Phys* 46, 115013.

Isaac, N., Saha, A.K., 2023. A review of the optimization strategies and methods used to locate hydrogen fuel refueling stations. *Energies (Basel)* 16, 2171.

Kume, T., Ikeda, Y., Iseki, T., Yakabe, H., Tanaka, H., Hikosaka, H., Takagi, Y., Ito, M., 2013. Performance evaluation of membrane on catalyst module for hydrogen production from natural gas. *Int J Hydrogen Energy* 38, 6079–6084.

Moral, G., Ortiz, A., Gorri, D., Ortiz, I., 2024. Hydrogen recovery from industrial waste streams using Matrimid®/ZIF mixed matrix membranes. *Int J Hydrogen Energy* 51, 210–224.

Seiiedhoseiny, M., Ghasemzadeh, K., Mohammadpourfard, M., 2024. The recovery of hydrogen from ammonia production processes. In *Progresses in Ammonia: Science Technology and Membranes*; Varma, A. K. S. G. Ed. Elsevier, pp. 21–42.

Singla, S., Shetti, N.P., Basu, S., Mondal, K., Aminabhavi, T.M., 2022. Hydrogen production technologies-membrane based separation, storage and challenges. *J Environ Manage* 302, 113963.

5 Future Perspectives and Case Studies

5.1 FUTURE PERSPECTIVES

5.1.1 Membrane Materials

Emerging Materials: The future of hydrogen membrane technology lies in the discovery and development of novel materials. Researchers are exploring advanced polymers, metal-organic frameworks (MOFs), and hybrid materials for membranes. The aim is to achieve a balance between high selectivity, permeability, and stability.

Nanotechnology Integration: Incorporating nanotechnology for precise control at the molecular level is anticipated. Nanocomposite membranes with engineered nanoparticles can enhance material properties, leading to improved performance and durability.

Bio-inspired Materials: Inspiration from nature, such as biomimetic materials, is gaining attention. Materials mimicking natural structures, like channels found in biological membranes, are being explored to enhance the efficiency and selectivity of hydrogen separation.

Current Landscape: Membrane materials are at the forefront of innovation in hydrogen separation technology. Current membranes predominantly employ polymers, ceramics, and metal alloys. However, ongoing research is pushing the boundaries of material science to enhance the selectivity, permeability, and stability of membranes for hydrogen separation applications.

Emerging Materials:

1. **Advanced Polymers:**
   - *Current Status:* Traditional polymers like polymeric membranes have been widely used for hydrogen separation.
   - *Future Prospects:* Ongoing research explores advanced polymer materials with tailored structures, such as co-polymers and hybrid materials. These innovations aim to improve both selectivity and permeability, making polymers more competitive with other membrane materials (Shao et al., 2009).

DOI: 10.1201/9781003590682-5

2. **Metal-Organic Frameworks (MOFs):**
   - *Current Status:* MOFs show promise in gas separation due to their high porosity and tunable structures.
   - *Future Prospects:* Researchers are actively designing MOFs specifically for hydrogen separation. Customizable frameworks and functionalization of MOFs offer the potential for highly selective membranes.
3. **Hybrid Membrane Materials:**
   - *Current Status:* Combinations of polymers, ceramics, and nanoparticles have been explored to enhance membrane properties.
   - *Future Prospects:* Innovations in hybrid materials involve precise combinations of various elements to create membranes with synergistic properties. The goal is to create materials that are stronger, more selective, and more durable than individual components.

Nanotechnology Integration:

1. **Nanocomposite Membranes:**
   - *Current Status:* Nanocomposite membranes incorporate nanoparticles into traditional membrane materials.
   - *Future Prospects:* Nanotechnology enables precise control over membrane structures at the nanoscale. Future membranes may utilize engineered nanoparticles for improved gas transport properties, stability, and resistance to impurities.
2. **Thin-Film Nanocomposites:**
   - *Current Status:* Thin-film composite membranes are widely used in gas separation applications.
   - *Future Prospects:* Researchers are exploring the integration of nanomaterials into thin-film composites to enhance mechanical strength, selectivity, and permeability. Advanced deposition techniques like atomic layer deposition (ALD) and chemical vapor deposition (CVD) contribute to controlled nanoscale fabrication (Iulianelli and Drioli, 2020).

Bio-inspired Materials:

1. **Biomimetic Membranes:**
   - *Current Status:* Biomimetic membranes mimic natural structures, such as aquaporins, for selective transport.
   - *Future Prospects:* Further research aims to develop membranes inspired by various biological structures that exhibit remarkable selectivity and efficiency. These materials could revolutionize hydrogen separation by learning from nature's design principles.
2. **Bio-based Polymers:**
   - *Current Status:* Exploration of bio-based polymers for membrane applications is gaining attention.
   - *Future Prospects:* Utilizing renewable and biodegradable materials in membrane fabrication aligns with sustainability goals. Future membranes

may incorporate bio-based polymers, contributing to a more environmentally friendly approach.

Challenges and Considerations:

1. **Scalability and Cost:**
   - Developing advanced materials must consider scalability and cost-effectiveness for industrial applications.
2. **Integration with Current Infrastructure:**
   - New materials should be designed to integrate seamlessly with existing hydrogen separation systems.
3. **Robustness and Longevity:**
   - Future membranes need to demonstrate robustness and longevity under varying operating conditions.
4. **Sustainability:**
   - There is a growing emphasis on developing materials and processes with minimal environmental impact.

Conclusion:
The future of membrane materials for hydrogen separation holds immense promise, with researchers exploring a spectrum of innovations from advanced polymers to nanocomposites and bio-inspired materials. These advancements aim not only to improve separation performance but also to address challenges related to scalability, cost, and environmental sustainability, paving the way for the next generation of hydrogen membrane technologies.

5.1.2 Thin-Film Composite Membranes

Advanced Fabrication Techniques: Future developments in thin-film composite membranes involve refining fabrication techniques. Techniques like ALD and CVD are expected to enable precise control over film thickness, enhancing performance.

Multilayered Structures: Researchers are investigating multilayered composite structures to optimize selectivity and permeability. These designs aim to address the trade-off between selectivity and permeability, providing membranes tailored for specific applications.

Current Landscape: Thin-film composite (TFC) membranes have become integral to various separation processes, including hydrogen separation. These membranes typically consist of a thin selective layer atop a porous support layer. The current landscape sees widespread use of TFC membranes, particularly in gas separation applications (Seyedpour et al., 2024).

Advanced Fabrication Techniques:

1. **Atomic Layer Deposition (ALD) and Chemical Vapor Deposition (CVD):**
   - *Current Status:* TFC membranes are traditionally fabricated using methods like interfacial polymerization.
   - *Future Prospects:* ALD and CVD are emerging as advanced techniques for membrane fabrication. These technologies offer precise control over film thickness and composition, enabling the production of TFC membranes with enhanced properties and performance.
2. **Precision Engineering of Thin Films:**
   - *Current Status:* Thin films are crucial for membrane selectivity and permeability.
   - *Future Prospects:* Future research aims to engineer thin films at the nanoscale, utilizing advanced manufacturing techniques to control the arrangement of molecules. This level of precision can lead to improved gas separation performance.

Multilayered Structures:

1. **Selectivity-Permeability Optimization:**
   - *Current Status:* The trade-off between selectivity and permeability is a key consideration in TFC membranes.
   - *Future Prospects:* Multilayered composite structures are being explored to strike a balance between high selectivity and permeability. The design involves stacking different selective layers to optimize overall membrane performance for specific gas separation applications.
2. **Tailored Functionalities:**
   - *Current Status:* TFC membranes are designed for specific gas separations.
   - *Future Prospects:* Tailoring functionalities of individual layers within TFC membranes is a future focus. Researchers are exploring ways to incorporate selective functional groups that enhance the separation of target gases like hydrogen while maintaining stability.

Challenges and Considerations:

1. **Stability under Harsh Conditions:**
   - Achieving stability under high-temperature and high-pressure conditions remains a challenge, particularly for TFC membranes designed for industrial applications.
2. **Scaling up Production:**
   - Transitioning from laboratory-scale to large-scale production without compromising membrane performance poses a challenge that requires innovative solutions.

3. **Integration with Advanced Support Layers:**
    - Exploring advanced materials for support layers that enhance mechanical strength and overall stability is a priority.
4. **Enhancing Resistance to Impurities:**
    - TFC membranes need to be designed to resist degradation in the presence of impurities commonly found in industrial gas streams.

Conclusion:
TFC membranes hold great potential for advancing hydrogen separation technology. Future developments focus on employing cutting-edge fabrication techniques like ALD and CVD, optimizing multilayered structures, and tailoring functionalities for improved selectivity and permeability. Overcoming challenges related to stability, scalability, and impurity resistance will be critical in realizing the full potential of advanced TFC membranes in diverse industrial applications.

5.1.3 High-Temperature Hydrogen Membranes

Advanced Ceramics and Alloys: High-temperature hydrogen membranes are envisioned to play a crucial role in industrial processes. The focus is on developing advanced ceramics and alloys that can withstand elevated temperatures, expanding the range of applications in industries like steel manufacturing and syngas production.

Integrated Heat Management: Incorporating integrated heat management systems will be a key aspect. Research is directed toward designing membranes that can operate at high temperatures while efficiently managing heat, contributing to overall system efficiency.

Current Landscape: High-temperature hydrogen separation membranes represent a crucial area of development in membrane technology. These membranes are designed to operate in environments with elevated temperatures, offering unique advantages for various industrial processes (Gallucci et al., 2017).

Advanced Ceramics and Alloys:

1. **Ceramic Membranes:**
    - *Current Status:* Ceramic materials like zeolites and silica-based membranes are known for their stability at high temperatures.
    - *Future Prospects:* Advancements involve the exploration of novel ceramic materials with improved hydrogen permeation properties and resistance to thermal stresses.
2. **High-Temperature Alloys:**
    - *Current Status:* Certain metallic alloys demonstrate high-temperature resistance.
    - *Future Prospects:* Research is focused on alloy compositions that withstand extreme temperatures while maintaining structural integrity for

prolonged periods, especially in applications like syngas production and steel manufacturing.

Integrated Heat Management:

1. **Thermal Stability:**
   - *Current Status:* High temperatures can impact membrane stability and performance.
   - *Future Prospects:* Innovative heat management systems, including insulation and active temperature control, are being integrated to ensure optimal membrane performance while preventing degradation.
2. **Thermal Cycling Considerations:**
   - *Current Status:* Thermal cycling can affect membrane durability.
   - *Future Prospects:* Future membranes will be designed to withstand frequent thermal cycling, a common occurrence in industrial processes. Materials with minimal thermal expansion and contraction will be explored.

Challenges and Considerations:

1. **Mechanical Integrity**:
   - Ensuring that high-temperature membranes maintain mechanical strength and structural integrity over time is a critical challenge.
2. **Hydrogen Embrittlement:**
   - High temperatures can exacerbate hydrogen embrittlement issues in certain materials.
   - *Mitigation Strategies:* Developing alloys and ceramics with inherent resistance to hydrogen embrittlement is crucial. Incorporating preventive measures, such as surface coatings or alloy modifications, will be explored.
3. **Industrial Applicability:**
   - Integrating high-temperature membranes into existing industrial processes poses challenges.
   - *Innovative Solutions:* Collaborative efforts between membrane researchers and industry experts will be essential to design membranes that seamlessly integrate into high-temperature applications.

Conclusion:

The future of high-temperature hydrogen membranes is marked by advancements in ceramic and alloy materials, as well as the integration of sophisticated heat management systems. Overcoming challenges related to mechanical integrity, hydrogen embrittlement, and seamless integration into industrial processes will be key in realizing the full potential of these membranes for applications in syngas production, steel manufacturing, and other high-temperature environments.

5.1.4 Membrane Reactors

Hybrid Systems Integration: The future of hydrogen membrane reactors involves the integration of multiple technologies. Combining membrane reactors with catalytic processes can enhance reaction kinetics, improve yields, and simplify downstream separation processes.

Advanced Catalytic Materials: Advancements in catalytic materials compatible with membrane reactors are anticipated. Novel catalysts that promote hydrogen production and selectively enhance reaction rates are under exploration for improved overall reactor efficiency.

Current Landscape: Membrane reactors represent an innovative approach where membranes are integrated directly into the reaction vessel, facilitating simultaneous reaction and separation. In the context of hydrogen production, membrane reactors hold immense potential for improving reaction kinetics, enhancing yields, and simplifying downstream separation processes (Gallucci et al., 2017).

Hybrid Systems Integration:

1. **Synergistic Integration of Membrane and Catalytic Processes:**
    - *Current Status:* Membrane reactors are integrated with various catalytic processes for hydrogen production.
    - *Future Prospects:* Future advancements involve optimizing the synergy between membranes and catalysts to improve overall reactor efficiency. This includes tailoring membrane properties to complement specific catalytic reactions.
2. **Dual-Functionality Membranes:**
    - *Current Status:* Membranes selectively allow certain components to permeate while inhibiting others.
    - *Future Prospects:* The development of dual-functionality membranes that simultaneously facilitate hydrogen separation and promote specific catalytic reactions is an area of active research. These membranes act as catalyst supports, enhancing reaction rates.

Advanced Catalytic Materials:

1. **Tailored Catalysts for Membrane Reactors:**
    - *Current Status:* Catalysts play a pivotal role in membrane reactor performance.
    - *Future Prospects:* Future research focuses on tailoring catalysts specifically for membrane reactor applications. This involves designing catalysts that work synergistically with membranes to enhance reaction rates and selectivity.
2. **Immobilized Catalysts:**
    - *Current Status:* Catalyst immobilization is a common approach in membrane reactors.

- *Future Prospects:* Innovations involve developing advanced methods for catalyst immobilization, ensuring stable and efficient catalyst performance over extended operational periods.

Challenges and Considerations:

1. **Selectivity-Permeability Balance:**
   - Achieving an optimal balance between membrane selectivity and permeability is a challenge, particularly when integrating with specific catalytic processes.
2. **Longevity of Catalysts:**
   - Ensuring the longevity and stability of catalysts within membrane reactors is crucial for sustained performance.
3. **Scale-Up Challenges:**
   - Transitioning from laboratory-scale membrane reactors to industrial-scale applications requires addressing scale-up challenges effectively.
4. **Process Intensification:**
   - Maximizing the benefits of membrane reactors in terms of process intensification while minimizing energy consumption is an ongoing consideration.

Conclusion:
The future of membrane reactors in hydrogen production envisions tighter integration with catalytic processes, resulting in more efficient and streamlined reaction systems. Advances in tailoring catalysts for membrane reactors, optimizing dual-functionality membranes, and addressing scale-up challenges are pivotal for realizing the full potential of these reactors in industrial applications. As research progresses, membrane reactors are likely to play a central role in sustainable and energy-efficient hydrogen production processes.

5.1.5 Hydrogen Economy

Large-Scale Integration: As the hydrogen economy expands, hydrogen membrane technology is expected to play a pivotal role in large-scale hydrogen production, storage, and distribution. Integrated systems with renewable energy sources for green hydrogen production will become more prevalent.

Hydrogen Storage Technologies: Innovations in hydrogen storage technologies, including membrane-based storage systems, are foreseen. Efficient and safe hydrogen storage methods are essential for realizing the full potential of the hydrogen economy.

Current Landscape: The hydrogen economy is a comprehensive concept that envisions hydrogen as a clean and versatile energy carrier, with applications spanning various sectors, including transportation, industry, and power generation. The development of a robust hydrogen economy relies on advancements in production, storage, and utilization technologies (Tseng et al., 2005).

Large-Scale Integration:

1. **Green Hydrogen Production:**
    - *Current Status:* Green hydrogen, produced through renewable methods like electrolysis, is gaining prominence.
    - *Future Prospects:* Large-scale integration involves expanding green hydrogen production capacity, utilizing renewable energy sources such as wind and solar to meet the growing demand.
2. **Blue Hydrogen Production:**
    - *Current Status:* Blue hydrogen, produced from natural gas with carbon capture and storage (CCS), is part of the current hydrogen landscape.
    - *Future Prospects:* Ongoing research focuses on improving the efficiency of carbon capture technologies and optimizing the overall environmental footprint of blue hydrogen production.

Hydrogen Storage Technologies:

1. **Advanced Storage Materials:**
    - *Current Status:* Hydrogen storage methods include compressed hydrogen and liquid hydrogen.
    - *Future Prospects:* Advancements involve the exploration of novel materials for solid-state hydrogen storage, such as metal hydrides and chemical hydrogen storage, to enhance safety and energy density.
2. **Infrastructure Development:**
    - *Current Status:* Developing hydrogen infrastructure, including storage facilities and transportation networks, is underway.
    - *Future Prospects:* Continued efforts in infrastructure development are crucial, emphasizing the need for standardized storage and transportation systems to support the widespread adoption of hydrogen.

Challenges and Considerations:

1. **Cost Reduction Strategies:**
    - Achieving cost competitiveness with other energy carriers is a significant challenge for the hydrogen economy.
    - *Mitigation Strategies:* Ongoing efforts include optimizing production methods, reducing equipment costs, and improving overall system efficiency.
2. **Integration with Renewable Sources:**
    - Seamless integration with renewable energy sources is essential for realizing the environmental benefits of the hydrogen economy.
    - *Technological Integration:* Integrating hydrogen production with intermittent renewable sources and developing energy storage solutions will be critical.

3. **Standardization and Regulation:**
    - Standardizing production, storage, and transportation methods is essential for ensuring safety and interoperability in the hydrogen economy.
    - *Regulatory Frameworks:* Governments and international organizations play a crucial role in developing regulatory frameworks that promote standardization and safety.

Conclusion:
The future of the hydrogen economy revolves around increasing the share of green hydrogen, advancing storage technologies, and establishing a robust infrastructure. Overcoming challenges related to cost, integration with renewable sources, and standardization will be pivotal for the widespread adoption of hydrogen as a key player in the global energy landscape. As research and development progress, the hydrogen economy is poised to contribute significantly to decarbonization efforts across various sectors.

5.1.6 Economic Analysis

Cost Reduction Strategies: Future economic analyses will focus on identifying cost reduction strategies. Continuous advancements in materials, manufacturing processes, and system integration are expected to contribute to lowering the overall costs of hydrogen membrane technologies.

LCA and Techno-Economic Assessments: In-depth life cycle assessments (LCA) and techno-economic assessments will become standard practices. These analyses will provide comprehensive insights into the economic and environmental viability of hydrogen membrane applications.

Current Landscape: Economic analysis plays a crucial role in determining the viability and competitiveness of hydrogen membrane technologies. The current landscape involves assessing the costs associated with membrane production, installation, and operation, as well as comparing these with alternative hydrogen production and separation methods.

Cost Reduction Strategies:

1. **Material Costs:**
    - *Current Status:* Membrane material costs contribute significantly to overall system expenses.
    - *Future Prospects:* Innovations in material science, process optimization, and scalable production methods aim to reduce material costs, enhancing the economic viability of hydrogen membrane technologies.
2. **Manufacturing Processes:**
    - *Current Status:* Manufacturing techniques impact the overall cost of membrane production.
    - *Future Prospects:* Advancements in manufacturing processes, including advanced deposition methods and automation, seek to streamline production, reduce labor costs, and improve overall efficiency.

Life Cycle Assessments (LCA) and Techno-Economic Analyses:

1. **Comprehensive Evaluation:**
   - *Current Status:* LCA and techno-economic analyses provide insights into the environmental and economic aspects of membrane technologies.
   - *Future Prospects:* Future analyses will likely involve more comprehensive assessments, considering the entire life cycle of hydrogen membrane systems and their integration into various industrial processes.
2. **Integration with Renewable Energy:**
   - *Current Status:* Evaluating the economic feasibility of integrating hydrogen membrane technologies with renewable energy sources is critical.
   - *Future Prospects:* As renewable energy becomes more prevalent, economic analyses will need to account for the dynamic nature of energy costs and the role of hydrogen membrane technologies in supporting renewable energy integration (Salkuyeh et al., 2017).

Challenges and Considerations:

1. **Commercial Viability:**
   - *Challenge:* Achieving commercial viability is essential for the widespread adoption of hydrogen membrane technologies.
   - *Strategies:* Continued research, technological advancements, and collaboration between academia and industry are vital for enhancing the commercial viability of membrane-based hydrogen processes.
2. **Incentives and Subsidies:**
   - *Challenge:* The economic landscape is influenced by government incentives and subsidies.
   - *Considerations:* Future economic analyses must consider the evolving landscape of policy support, including subsidies and incentives that impact the economic competitiveness of hydrogen membrane technologies.

Conclusion:
The future of economic analysis in the context of hydrogen membrane technologies involves addressing challenges related to cost reduction, conducting more comprehensive life cycle assessments, and adapting to evolving energy landscapes. By optimizing material costs, manufacturing processes, and integrating with renewable energy sources, hydrogen membrane technologies can become economically competitive, fostering their widespread adoption and contributing to a sustainable and economically viable hydrogen ecosystem.

5.1.7 Environmental Impact

Green Manufacturing Practices: The future emphasizes the adoption of green manufacturing practices for membrane production. Researchers are exploring environmentally friendly synthesis methods and recyclable materials to reduce the environmental footprint of membrane technologies.

Carbon Capture and Utilization: Hydrogen membrane systems are anticipated to contribute to carbon capture and utilization efforts. Integration with processes for capturing and utilizing carbon emissions will enhance the environmental sustainability of hydrogen production.

Current Landscape: Assessing the environmental impact of hydrogen membrane technologies is crucial for ensuring sustainability. The current landscape involves evaluating factors such as energy consumption, carbon footprint, and resource utilization associated with membrane production, operation, and disposal (Nowotny and Veziroglu, 2011).

Green Manufacturing Practices:

1. **Sustainable Materials:**
   - *Current Status:* Materials used in membrane production contribute to the environmental footprint.
   - *Future Prospects:* The future envisions the adoption of sustainable materials and greener manufacturing practices, including the use of recyclable materials and environmentally friendly synthesis methods.
2. **Energy-Efficient Processes:**
   - *Current Status:* The energy consumption during membrane production and operation is a consideration.
   - *Future Prospects:* Advancements in energy-efficient manufacturing processes and operational strategies aim to reduce the overall environmental impact of hydrogen membrane technologies.

Carbon Capture and Utilization:

1. **Integration with Carbon Capture:**
   - *Current Status:* Addressing carbon emissions associated with hydrogen production is crucial.
   - *Future Prospects:* Future hydrogen membrane technologies are likely to integrate with carbon capture and utilization strategies, contributing to overall carbon reduction in industrial processes.
2. **Life Cycle Assessments (LCAs):**
   - *Current Status:* LCA helps evaluate the environmental impact of membrane technologies.
   - *Future Prospects:* Continuous refinement of LCA methodologies and expanding assessments to cover the entire life cycle of hydrogen membranes will provide more accurate insights into their environmental footprint.

Challenges and Considerations:

1. **Resource Scarcity:**
   - *Challenge:* Some membrane materials may rely on scarce resources, impacting their overall sustainability.
   - *Considerations:* Future research will explore alternative materials and recycling strategies to mitigate the environmental impact of resource scarcity.
2. **End-of-Life Considerations:**
   - *Challenge:* Proper disposal and recycling of membranes at the end of their life cycle pose challenges.
   - *Considerations:* Implementing efficient recycling methods and exploring materials with minimal environmental impact at disposal will be important considerations.

Conclusion:
The future of hydrogen membrane technologies involves a strong commitment to minimizing their environmental impact. By adopting green manufacturing practices, integrating with carbon capture strategies, and continually improving life cycle assessments, these technologies can contribute to a more sustainable and environmentally friendly hydrogen production landscape. Addressing challenges related to resource scarcity and end-of-life considerations will be essential for realizing the full potential of environmentally conscious hydrogen membrane technologies.

5.2 CASE STUDIES

5.2.1 Hydrogen Separation in Petrochemical Industry

Background:
In this case study, a petrochemical facility aimed to optimize its hydrogen recovery process. The facility, engaged in various petrochemical processes, generates hydrogen as a byproduct. The challenge was to enhance the purity of the recovered hydrogen for use within the facility and reduce reliance on traditional purification methods.

Implementation:
The facility integrated a hydrogen membrane separation system into its existing processes. Polymer-based membranes were selected for their high selectivity, allowing hydrogen to permeate through while excluding impurities such as methane and other hydrocarbons. The membranes were configured to operate under the specific temperature and pressure conditions of the petrochemical processes.

Results:

1. **Increased Hydrogen Purity:**
   - The membrane separation system significantly increased the purity of the recovered hydrogen. By selectively allowing only hydrogen molecules to pass through the membrane, impurities were effectively removed.

- This high-purity hydrogen is crucial for maintaining the quality and efficiency of downstream petrochemical processes.

2. **Operational Efficiency:**
   - The continuous operation of the membrane system contributed to improved overall efficiency. Unlike traditional purification methods that might involve complex and energy-intensive processes, the membrane system provided a streamlined and energy-efficient solution.
   - Operational efficiency gains were realized as the facility could rely on the membrane system for on-site hydrogen purification.

Conclusion:
The implementation of hydrogen membrane separation technology in the petrochemical facility proved to be a success. The use of polymer-based membranes tailored to the specific needs of the petrochemical industry resulted in a substantial increase in hydrogen purity and operational efficiency. This case study demonstrates the practical application of hydrogen membrane technologies in industrial settings, showcasing their ability to address specific challenges and contribute to the optimization of hydrogen recovery processes in the petrochemical sector.

5.2.2 Hydrogen Recovery in Ammonia Production

Objective: Enhancing hydrogen recovery efficiency in ammonia production.

Background:
In this case study, an ammonia production plant sought to enhance its hydrogen recovery process. Hydrogen is a critical component in ammonia synthesis, and traditional recovery methods were identified as resource intensive. The plant aimed to improve efficiency, increase hydrogen yield, and align its processes with sustainability goals.

Implementation:
The plant introduced a hydrogen membrane system into its hydrogen recovery process. The selected membranes were high-temperature ceramic membranes designed to withstand the conditions of ammonia production. The ceramic membranes were integrated into the system to selectively allow hydrogen to permeate through while excluding other gases.

Results:

1. **Enhanced Hydrogen Yield:**
   - The introduction of the ceramic membrane system led to a notable increase in the yield of recovered hydrogen. The selective permeation of hydrogen through the membranes improved the overall efficiency of the recovery process.
   - The enhanced yield had a direct positive impact on the ammonia production process, where hydrogen is a crucial feedstock.

2. **Energy Savings:**
    - The membrane system demonstrated energy savings compared to conventional recovery methods. The optimized recovery process contributed to the plant's efforts to reduce its overall energy consumption and environmental footprint.
    - The energy-efficient operation of the membrane system aligned with the plant's sustainability goals.

Conclusion:
The implementation of a high-temperature ceramic membrane system in the ammonia production plant showcased the effectiveness of hydrogen membrane technologies in an industrial context. The increased hydrogen yield and energy savings demonstrated the potential for these technologies to optimize ammonia production processes. This case study emphasizes the adaptability of membrane systems to specific industrial conditions and their role in contributing to both efficiency and sustainability objectives in ammonia synthesis.

5.2.3 Hydro Recovery from Biogas

Objective: Efficient hydrogen recovery from biogas generated at a wastewater treatment plant.

Background:
In this case study, a wastewater treatment plant with anaerobic digestion processes aimed to harness the hydrogen content in biogas for clean energy production. Biogas, a byproduct of organic waste treatment, contains hydrogen along with methane and other impurities. The challenge was to effectively recover and purify hydrogen from the biogas stream for utilization in various applications.

Implementation:
The wastewater treatment plant implemented a hydrogen membrane system designed to operate in the presence of impurities commonly found in biogas. Polymeric membranes were chosen for their resilience and selective permeability to hydrogen. The system was integrated into the biogas treatment process to recover hydrogen efficiently.

Results:

1. **Clean Hydrogen Production:**
    - The polymeric membrane system effectively separated and purified hydrogen from the biogas stream. By selectively allowing hydrogen molecules to permeate through the membranes, impurities were excluded, resulting in clean hydrogen.
    - The clean hydrogen produced was suitable for various applications, including energy generation and industrial processes.

2. **Waste-to-Energy Integration:**
   - The implementation of the hydrogen membrane system demonstrated the potential for integrating hydrogen recovery with waste-to-energy processes. By utilizing the hydrogen content in biogas, the wastewater treatment plant embraced a circular and sustainable approach to energy production.
   - The recovered hydrogen contributed to the overall energy output of the plant.

Conclusion:

The wastewater treatment plant's case study exemplifies the practical application of hydrogen membrane technologies in the context of waste-to-energy conversion. The use of polymeric membranes tailored for biogas conditions allowed for efficient hydrogen recovery, showcasing the adaptability of membrane systems to diverse industrial settings. This case study highlights the role of hydrogen membrane technologies in promoting sustainability by extracting valuable energy resources from waste streams.

REFERENCES

Gallucci, F., Medrano, J.A., Fernandez, E., Melendez, J., Van Sint Annaland, M., Pacheco-Tanaka, D.A., 2017. Advances on high temperature Pd-based membranes and membrane reactors for hydrogen purification and production. *J Membr Sci Res* 3, 142–156.

Iulianelli, A., Drioli, E., 2020. Membrane engineering: latest advancements in gas separation and pre-treatment processes, petrochemical industry and refinery, and future perspectives in emerging applications. *Fuel Process Technol* 206, 106464.

Nowotny, J., Veziroglu, T.N., 2011. Impact of hydrogen on the environment. *Int J Hydrogen Energy* 36, 13218–13224.

Salkuyeh, Y.K., Saville, B.A., MacLean, H.L., 2017. Techno-economic analysis and life cycle assessment of hydrogen production from natural gas using current and emerging technologies. *Int J Hydrogen Energy* 42, 18894–18909.

Seyedpour, F., Farahbakhsh, J., Dabaghian, Z., Suwaileh, W., Zargar, M., Rahimpour, A., Sadrzadeh, M., Ulbricht, M., Mansourpanah, Y., 2024. Advances and challenges in tailoring antibacterial polyamide thin film composite membranes for water treatment and desalination: a critical review. *Desalination* 581, 117614.

Shao, L., Low, B.T., Chung, T.-S., Greenberg, A.R., 2009. Polymeric membranes for the hydrogen economy: contemporary approaches and prospects for the future. *J Memb Sci* 327, 18–31.

Tseng, P., Lee, J., Friley, P., 2005. A hydrogen economy: opportunities and challenges. *Energy* 30, 2703–2720.

Index

I

L

M

N

O

P

R

S

T

W

Z